克里斯蒂安·德·鲍赞巴克

致谢

作者特别感谢卡莱·马丁（Gaelle Martin）的帮助、支持和她表现出的经久不变的幽默感。出于类似原因，在此也对弗朗索瓦·卡萨诺瓦（Françoise Casanova），奥雷利·兰博瑞（Aurélie Lambrecht），卢·蒙哈德（Lou Mollgaard），艾蒂西亚·维西奥特（Laetitia Viciot），加布里拉·威尔逊－桑德勒（Gabriela Wilson-Sadler）和依莲娜·布劳姆（Hélène Brom），吉蒂·达格（Gitty Darugar），德蒂·冯·沙文（Deidi von Schaewen）和艾蒂安·皮尔斯（Étienne Pierrès）一并致谢。

Acknowledgements

The author would particularly like to thank Gaelle Martin for her assistance, support and unfailing good humor, as well as Françoise Casanova, Aurélie Lambrecht, Lou Mollgaard, Laetitia Viciot, Gabriela Wilson-Sadler and Hélène Brom, Gitty Darugar, Deidi von Schaewen and Étienne Pierrès for much the same reasons.

à/for Églée, Aïna et Basile

克里斯蒂安·德·鲍赞巴克

Christian de Portzamparc

[法] 吉勒斯·德·比尔　著
王建武　译

中国建筑工业出版社

著作权合同登记图字：01－2007－3725 号

图书在版编目（CIP）数据

克里斯蒂安·德·鲍赞巴克 /（法）比尔著；王建武译．—北京：中国建筑工业出版社，2010
ISBN 978－7－112－11805－2

Ⅰ．克… Ⅱ．①比…②王… Ⅲ．建筑设计－作品集－法国－现代 Ⅳ．TU206

中国版本图书馆 CIP 数据核字（2010）第 028748 号

责任编辑：姚丹宁
责任设计：郑秋菊
责任校对：刘　钰

克里斯蒂安·德·鲍赞巴克
Christian de Portzamparc
[法] 吉勒斯·德·比尔　著
王建武　译

*
中国建筑工业出版社出版、发行（北京西郊百万庄）
各地新华书店、建筑书店经销
北京嘉泰利德公司制版
北京中科印刷有限公司印刷
*
开本：787×960 毫米　1/16　印张：19　字数：365 千字
2010 年 7 月第一版　2010 年 7 月第一次印刷
定价：78.00 元
ISBN 978－7－112－11805－2
(19032)

1
2
3
4
5

6
7
8
9
10

1

现实情形
existing

生活之道
The business of living

以下是克里斯蒂安·德·鲍赞巴克与法国作家菲利普·索莱尔的谈话——（内容）海阔天空，涉及生活、思想、世事、爱情、感受、音乐、文学、成功、失败、旅行……说笑间，他开始谈论索莱尔的写作方式："我愿意探讨您小说中的形式问题。举例来说，我喜欢《顽念》（Passion Fixe）描述的混沌状态：文章通过反复重叠与复杂（的描写）使其形式趋向迷宫和万花筒似的效果。您自己曾将其与立体派绘画相比较。而我则很欣赏梦境、旭日、女性、广告、中国诗歌，以及政治、哲学、色情和讽刺性的部分。一切皆有存在的空间。在我看来，这部书还采用另一方式消除行文的直线性：即多个主题的相互穿插——至少 12 ~ 13 个主题相互交织，以不同的基调谈及生活、历史、隐私、叙述、自语、独占——（甚至）还包括性接触，其处理方式可说是另辟蹊

So here's Christian de Portzamparc chatting with French writer Philippe Sollers—about everything and nothing, about life, ideas, what's happening in the world, love, feelings, music, literature, success, failure, travel... All of a sudden he embarks on an analysis of Sollers' approach to writing: "I'd like to talk about form in your novels. I love the mad tangle of *Passion Fixe for* instance: you've got form veering off into a labyrinth or kaleidoscope, with all those overlays and intricacies and doubling back. You've compared it yourself to a cubist painting. I just love the dreams, the sunrises, the women, the advertisements, the Chinese poems and the political, philosophical, erotic and sarcastic bits. There's a moment there for everything. It seems to me that in this book there's yet another way of getting rid of textual linearity: you could almost call it a manual rather than a novel, with its themes intertwining—there must be twelve or thirteen of them that keep on crisscrossing, different tones intersecting with life, history,

克里斯蒂安·德·鲍赞巴克
Christian de Portzamparc

径。你甚至可以缘此将其作为一部便览手册而非小说。在这部破碎化的文本中，您（索莱尔）论及的不同主题相继而生并形成新的统一整体。"

"阅读《顽念》(Passion Fixe)，"鲍赞巴克继续说道，"我发现自己惊愕于形式与统一，为其中领略到的众多'岛屿'（指不同的主题——译者注）而动容，在您的小说中，'解体'找到了其文学表达。而这也是世界和城市的存在形式：重叠、碎片、通信、普遍存在、高速行进、多种媒介，解体与挫败……"他总结道："《顽念》的形式是破碎的，但一切在视觉上却保持常态，就像是有机且连续的陈述。如同生活一样，您会说，没有什么统治性因素，也与今天的城市异曲同工。您是否对重叠形式的问题感兴趣？从音乐的整体感维度，我的意思是，从对作品整体性的感知出发。"（选自 Voirécrire，Paris，Calmann-Lévy，2003 年）

这是鲍赞巴克头戴面纱的绝佳自画像！一切对他来说至关重要，能够解释、定义和追踪锁定他的词汇，像游行队伍似地整齐罗列起来：破碎、形式、整体、多样性、解体、重叠、有机的连续性、音乐感、感知。这一形象与雅克·拉肯（Lacan）的评论不谋而合："破碎导致整体的创造，整体中的各部分较其复杂的组织方式和辩证的相互关系为轻。从这一观点分析，克里斯蒂安·德·鲍赞巴克主要关注相互关系和外部体型的均衡，描绘和塑造乍看起来生硬武断的建筑主题，而实际上，这些主题是卓越建筑句法的组成元素，混合了纯粹的形式和历史记忆。鲍赞巴克并非意欲创造巨大而孤立的结构体，而是要勾勒出具有不对称性和极性的统一整体，他称之为对现代建筑的善意批判。"

"克里斯蒂安·德·鲍赞巴克的志趣在于为业已建立的事物——情境与间隔、空旷地与图腾柱——加入二元性：象征的

"无名诗人墓"：到达巴黎美术学院后的第一幅草图，1963 年。
"Tomb for an Unknown Poet": his first sketch on arriving at the Ecole des Beaux-Arts in Paris, 1963.

鲍赞巴克著名的铅笔线条。无题，铅笔，1976 年。
Portzamparc's celebrated pencil strokes. Untitled, pencil on paper, 1976.

intimacy, narrative, solitude conversation and exclusive possession — and with making love, which is treated in a really unique way. In this fragmented text, Sollers' different themes follow on from each other and form new wholes.

"Reading *Passion Fixe*," Portzamparc goes on, "I find myself wondering about form and unity. All those islands the book takes you to. In your novels breakup actually finds literary form. Because it's also the form of the world, and the city: overlays, fragmentations, telecommunications, ubiquity, high-speed travel and multiple media. Breakup and frustration..." And he winds up: "The form of Passion Fixe is broken up, but visually everything looks normal, as if this were an organically continuous narrative. Like life, you'll reply, which has no dominant form. Which is exactly like the city of today. Does the question of *overarching form* interest you? In the musical sense, I mean, in the sense of a perception of the whole of a construction, of its unity ." (In *Voir-Écrire*, Paris, Calmann -Lévy, 2003)

What a splendidly veiled self-portrait of Portzamparc himself! All the words that are important to him, describe him, define him, home in on him, are lined up as if on parade: fragmentation, form, unity, multiplicity, breakup, overlay, organic continuity, musicality, perception. A portrait that fits readily with the one offered by Jacques Lacan: "The fragmentation problem leads to the creation of a whole whose component parts are less important than their complex organization and the dialectic of their relationships. Seen in this light Christian de Portzamparc focuses on relationships and the proper balance of external volumes, drawing on architectural motifs which at first glance seem arbitrary but which in fact are elements of a singular architectural syntax mingling pure form and historical reminiscence. De Portzamparc is not out to create a monolithic structure, but rather to outline a unitary whole whose multiple dissymmetries and polarities radiate what he calls a loving critique of modern architecture. "

和功利的，联系的和主观的，沉重与轻盈，图形与基底……”（选自法国建筑，1940-2000 年，巴黎，Le Mouiteur, 2001 年）（Architecture en France,1940-2000 年，Paris, Le Mouiteur, 2001 年）

鲍赞巴克与索莱尔之间的关联是什么？或许是二者身上皆存的 18 世纪法国（精神）的余绪。不是 18 世纪的浪漫主义，而是那个时代弥足珍贵的特征——明朗、清晰性，以及独立自主精神的觉醒和对抒情诗式幻想的抵触。索莱尔或许如是——而马里沃（Marivaux）则毫无疑问（抱有该思想）。

这是否意味着鲍赞巴克仅能通过文学才能被了解？通过文学的、辩证法、体裁上的策略？采用（特有）的建筑模式，他的建筑特征与风格是否兼备大众化气质与博学性、挽歌体情调与辛辣的讽刺性、隐喻重重而又生动具体、华美奢侈却又不失深沉？

1996 年蓬皮杜中心举办了鲍赞巴克（设计成果）展——“工作室纵览”（Studio Scenes），索莱尔作的目录序言拥有颇具启发性的标题——“作为思想的建筑”，序言中写道：“我很高兴鲍赞巴克为建筑而写作，也为现实的影响，为表现某些阴暗放纵的内在状态的吸引力与暗示作用而写作。这是小说家和诗人的话语模式。场所创造和文字组织肇始于相同的源。”正如约瑟夫·布洛茨基曾向密友所罗门·沃克沃倾诉的那样：“任何对文学感兴趣的人都会或多或少地逾越规矩”——布洛茨基的古典主义风格并未有悖于他对俄国语言的根本性改变。

“Christian de Portzamparc's pleasure lies in adding other dualities to those — situation and spacing, the clearing and the totem pole — already established: the symbolic and the utilitarian, the relational and the subjective, heaviness and lightening, figure and background...” (In *Architecture en France*, 1940-2000, Paris, Le Moniteur, 2001)

So what's the connection between Sollers and Portzamparc? Probably that residue of 18th-century France you find in both. Not the 18th century Romanticism, but the 18th century of nimble lucidity, of sovereign disenchantment, of resistance to the lyrical illusion. Sollers probably — and Marivaux beyond a doubt.

Does this mean Portzamparc can only be approached through literature? Through literature, dialectics and stylistic devices? By taking his architectural mode, his manner and his style as simultaneously erudite and popular, elegiac and heavily ironic, metaphorical and concrete, sumptuous and sober?

As it happens, in “Architecture as Thought”, his luminously titled catalogue preface to *Scènes d'Atelier* (“Studio Scenes”), the Portzamparc exhibition at the Centre Pompidou in 1996, Philippe Sollers wrote, “I'm glad Christian de Portzamparc writes about architecture, about the impact of presence, about magnetism and intimations of some darkly extravagant interiority. This is the way poets and novelists talk. Place-creation and word-assembling proceed from the same basic source.” And Joseph Brodsky, whose classicism failed to prevent him from revolutionizing the Russian language, confided to his friend Salomon Volkov that “Anyone interested in literature knows he's more or less in breach of the law.”

建筑师玛梅在突尼斯设计的私人住宅，1998 年，鲍赞巴克进行改建、增建。

Private house built by the architect Marmey at Sidi Bousaid, in Tunisia. In 1998 Portzamparc extended and improved it.

鲍赞巴克从未停止过绘画、素描与雕塑，无论在家，在工作室还是在旅途中，他全年笔耕不辍。在“工作室纵览”进行之际，他解释这一强制性做法的目的是“对某些现象进行空间和造型探索，例如物质的伸展，空虚与盈满相互影响（的问题），再如对形式与内容的感知（问题）等。”他还补充说，他热衷于“即兴创作、无意识表演、现场发现等”。在此有必要进一步强调，在很大程度上，蓬皮杜中心举行的展览对鲍赞巴克本人也是有益和必要的。它不仅全面展示了设计师的作品及其多样性，而且对建筑师本人来说，也成为“自我解析”的一种形式。

在《读·写》(Voir-Ecrire) 中，他告诉

Portzamparc has never stopped painting, drawing, sculpting, all year round and whether at home, in the studio or traveling. At the time of the Scènes d'Atelier exhibition he explained that this compulsiveness was aimed at "spatial or plastic exploration of such phenomena as physical expansion, the interaction between void and fullness, and the perception of form and content." Adding that he is also given to "improvisations, acts born of impatience, and on-the-spot discoveries". At this point moreover, it is important to underline the extent to which the Centre Pompidou exhibition proved both useful and necessary for Portzamparc. Not only because it revealed the full extent and diversity of his work, but also because for the architect, it served as a form of "self-analysis."

绘画——全年笔耕不辍。
Drawing — all year round.

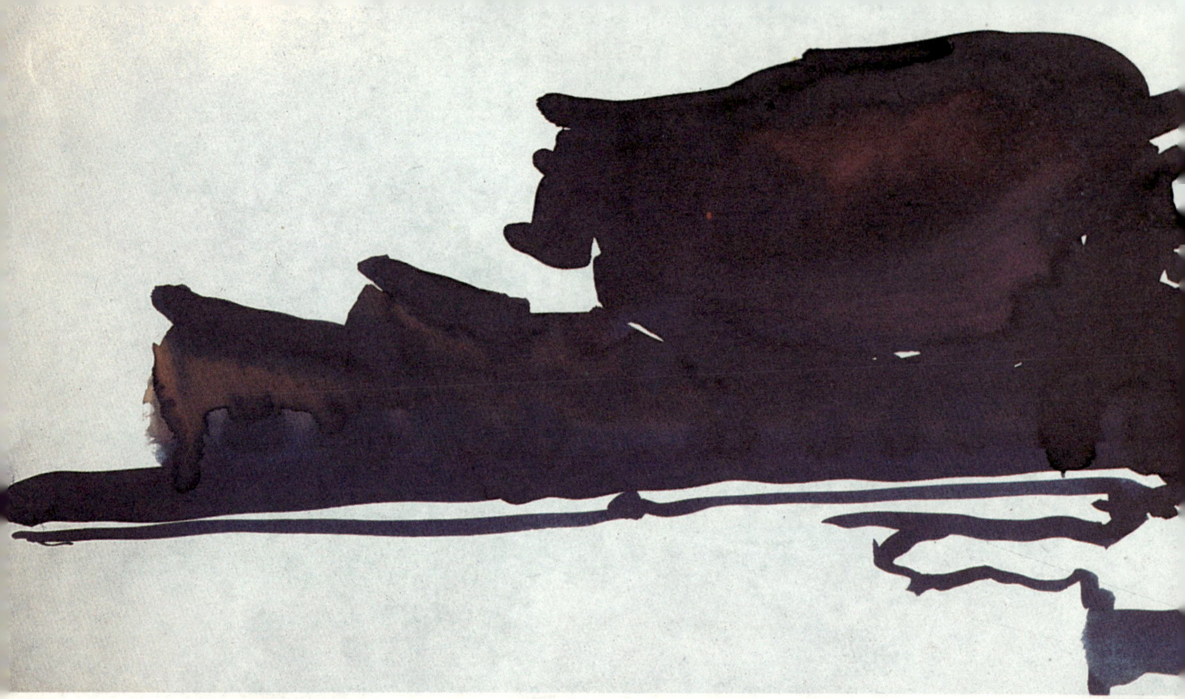

索莱尔，"在很长一段时间里，制图和语言对我来说完全互不相干。直到很晚的时候它们才开始彼此交融。此前，我的头脑分为两部分：一部分用于制图，另一部分用于写作。写作侵占了我大量时间：我完全沉浸其间，以至于放弃了设计工作——我不能一心二用。直到我已经设计了一些建筑，情况才变得容易处理一些。"

他的工作室同事也许会为那些无比精确的铅笔线条唏嘘不已，这些线条带给人延伸至无限的感觉，但这一工作方法同样招致不小的非议。在法国，那些没有继承传统，不能合于圭臬的事物——例如，身为建筑师，却同时从事素描、绘画、雕塑的人，只能被当作视觉艺术家。必须承认，在这个图像已经取代文字和线条的时代，

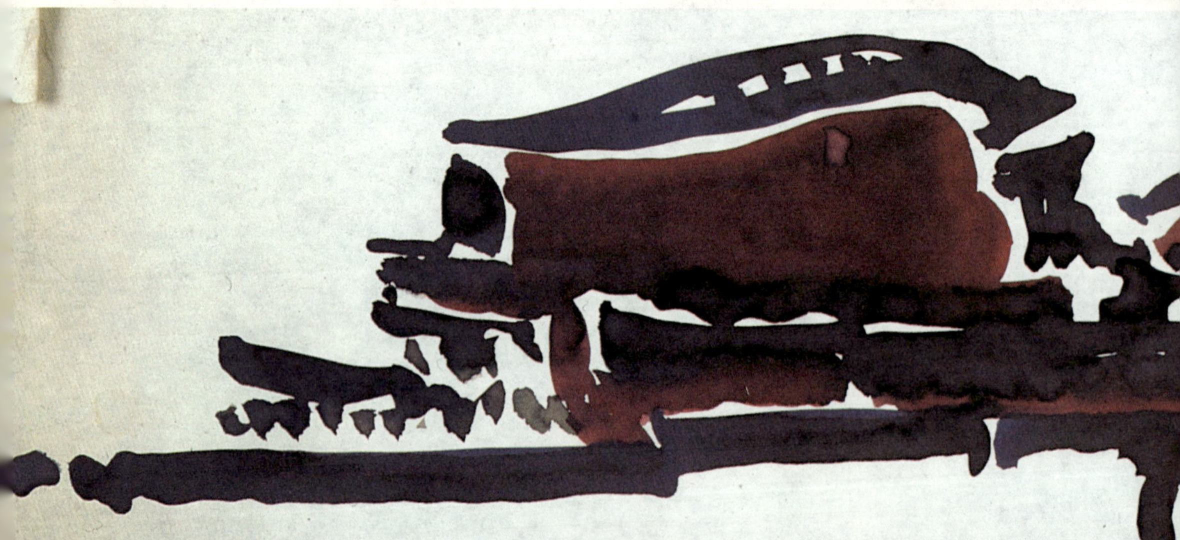

交响音乐厅方案，当时建筑师（鲍氏）仍在鲍顿因工作室学习，1963年。

Project for a Philharmonic Hall, when the architect was still a student at the Beaudoin atelier in 1963.

In *Voir-Ecrire* he tells Sollers, "For a long time drawing and language were completely separate for me. They only came together very late on. Before that I had two lobes in my brain: one drew, the other named. When I wrote it really ate into my time: I was so caught up in it I gave up working on projects — I couldn't handle writing and drawing at the same time. It was only after I had done some building that it became less complicated."

His studio associates might be knocked out by the sheer precision of those pencil strokes that can generate expansion ad infinitum, but this way of working has also attracted a lot of criticism. In France, where it doesn't go down well when things don't fit into the appropriate pigeonhole, an architect who draws, paints and sculpts just has to be a visual artist. Granted, his attitude is truly ambiguous in an age in which image has replaced word and line; yet Portzamparc is

鲍赞巴克的态度的确是含混的；但他机敏地意识到时代的变迁，而形式方面的探究并没有成为他在当代立足的绊脚石。

让我们回到文学，胡安·曼纽尔·德·布拉达曾宣称："任何文化制品都必须古老而又现代。普鲁斯特是最好的榜样，他汲取、吸收传统，然后从中孕育出全新的东西。"对鲍赞巴克作品的纵览让人联想到斯特拉文斯基的《普尔钦奈拉》与《春之祭》，毕加索的《亚威农的少女》与立体派，埃利尔·沙里宁赫尔辛基的巴洛克（建筑）与印第安纳哥伦布的现代主义教堂。对鲍赞巴克而言，融入历史洪流与推动历史前行是不可分割、不证自明的（两方面）。他的建筑作品中满怀特征鲜明、水乳交融的幻象与错觉，它们将可能性变成现实，不禁使人联想起教堂。

鲍赞巴克的"视觉艺术家"角色常常使他被人称为形式主义者。的确，对他来说，形式在塑造感知和形成基本的现象学方面至关重要；但在现实中，他更像游牧者，持续努力、不知疲倦地将他提出的主题拓展至更加广阔的领域：间隙、椭圆、圆锥、开口、尖拱、竖向倾斜、穿透的体块、底座、立柱、人行天桥、反射、色彩、城市风景、交叉与悬浮等等。

这些风格上的偏好与发人深省的对地理和场地的关注并行不悖，正所谓"纽约市的纽约人，巴黎市的巴黎人，蒙特利尔的魁北克人"。每一场地呼唤特定的回应——不是文脉的，而是特定的。"要想产生意义，"他断言，"则不能固守意义本身，而要在场地上下功夫。"尊重场所并且保持自我真实性实非易事，鲍赞巴克认为协调这两种不同趋势的关键在于复杂性，而非文脉主义。这一做法的成果逐渐

perfectly conscious of the movement of time and his formal commitment is no obstacle to being in the here and now.

To get back to literature, Juan Manuel de Prada has remarked that "Every cultural artifact must be simultaneously ancient and modern. Proust is the best example, in the way he digested tradition and built something utterly new out of it."

A bird's eye view of Portzamparc's work makes us think of Stravinsky moving from *Pulcinella* to the *Rite of Spring*, Picasso from the *Demoiselles d'Avignon* to Cubism, or Eliel Saarinen from his Helsinki Baroque to the Modernism of his church in Columbus, Indiana. For Portzamparc moving forward with history and taking history forward go self-evidently and inseparably together. Faced with the characteristic interplay of illusion that brings possibility into being in his architectural work, we can only think of cathedrals.

Portzamparc's "visual artist" side often leads to his being considered a formalist. True, form for him is capital in that it structures perception and shapes a fundamental phenomenology; but in reality he is much more of a nomad, playing on and tirelessly developing themes he extends towards ever more distant horizons: gaps, ellipses, cones, openings, pointed arches, vertical obliques, pierced blocks, pedestals, posts, footbridges, reflections, colors, views over the city, crossings and levitations.

These stylistic leanings are overlaid with a deeply felt taste for geography and the feeling of the site, for being "a New Yorker in New York, a Parisian in Paris, a Quebecker in Montreal". Each site calls for a specific response — not contextual, but specific. "What generates meaning," he asserts, "is working not on the meaning, but on the site." Respect for place while remaining true to oneself is no easy matter, so the issue for Portzamparc, faced with two different ways of being, is complexity, not contextuality. The upshot of this approach becomes clear when we realize that the Law

蒙特利尔，北京，里约热内卢：三个不同的方案，三种与众不同的形式，统一的风格。

Montreal, Beijing, Rio de Janeiro: three separate projects, three distinctive forms and true stylistic unity .

清晰起来。格拉斯法院，巴黎会议中心，布伦的“光亮的空间”（Espace Lumière），以及柏林的法国驻德大使馆等均在同一时间段设计，对于每个项目，鲍赞巴克均给与了特定回应，丝毫没有损害到自己的原则和出类拔萃的设计风格。他对地域及场地的直觉和敏感性使他每次都能发展和采纳英明、正确的构思。而1979年的欧风路（Hautes Formes）项目展示了被他称为“开放式街区”（理念）的适宜性与现代性。

主持马塞纳区的发展规划，使他有机会践行自己的城市发展三阶段理论：阶段一，阶层化城市；阶段二，规划的城市；阶段三，新时代（城市）框架的建构。“建筑被创造出来以缓解忧郁，”他低语道，引用费里波特·德洛尔姆的话。还有什么比这更具有怀旧情绪呢？

克里斯蒂安·德·鲍赞巴克1944年生于卡萨布兰卡，在他还不满一岁时，他的父亲——一名职业军人——回到业已解放的法国。后来的记录显示他们来往于法国与德国之间。

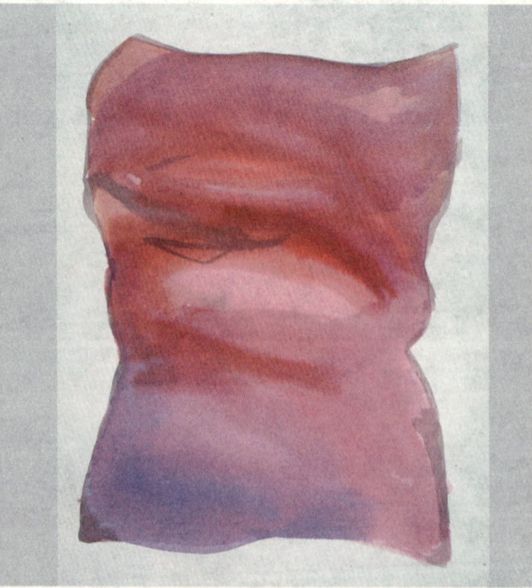

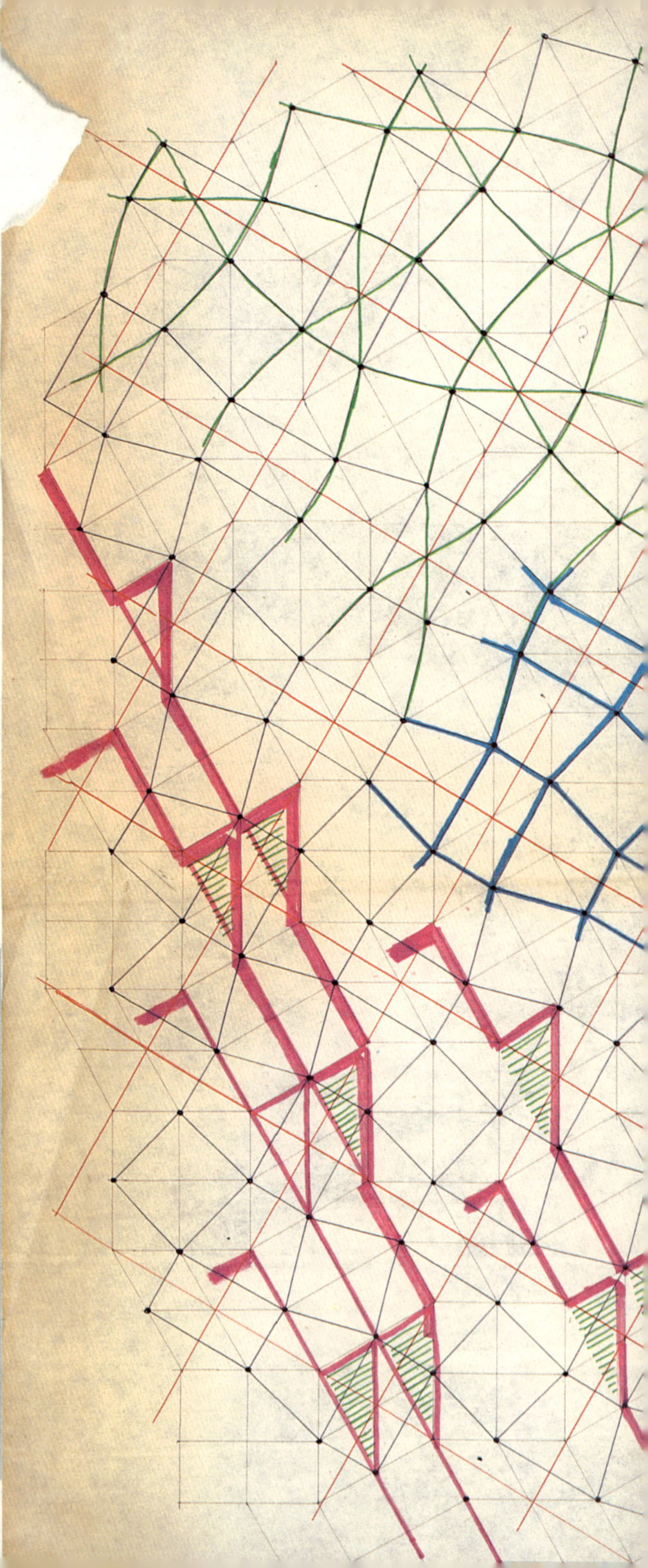

Courts in Grasse, the Convention Center in Paris, the Espace Lumière in Boulogne and the French Embassy in Berlin were all designed at the same time — and that each time Portzamparc provided a specific response without the least break with his principles or the least change to his distinctive style.

His taste and instinctive sensitivity to territory and site led Portzamparc to develop and apply ideas that turned out astonishingly right every time. With the Hautes Formes project in 1979 he demonstrated the aptness and modernity of what he called the "open block".

Being in charge of the development of the Masséna sector gave him the opportunity to implement his theory of the three ages of the city: Age 1, the stratified city; Age 2, the planned city; Age 3, the invention of new time frames.

"Architecture is made to soothe melancholy," he murmurs, quoting Philibert Delorme. What could be less nostalgic than that?

Christian de Portzamparc was born in Casablanca in 1944 and before he was a year old his father, a professional soldier, was back in liberated France. Subsequent postings saw them coming and going between France and Germany.

在坎迪利斯工作室设计的"非正交方案"。马克笔，1966 年。

Non-orthogonal student project at the Candilis atelier. Marker on paper, 1966.

（左页）无题，水彩，1975 年。

(On the left) Untitled, watercolor on paper, 1975

位于布列塔尼孔布尔城的家宅。
The family home near Combourg in Brittany .

鲍氏儿时同弟弟妹妹的合影。
As a child, with his younger brothers and sisters.

大海的召唤，无垠的地平线。
The sea beckons, with its infinite horizons.

可以确定，他们曾在孔布尔城外的家宅内度假，这里位于夏多布里昂地区，在圣马罗附近。这座庄园建于16世纪，在18世纪曾经改建，毋庸置疑，周围矗立的岩石，史前的墓葬坟冢，树木和池塘对建筑师产生过深刻影响，并且塑造、拓展了他的视野。随着父亲回归平民生活，他们举家搬入雷恩，克里斯蒂安开始了新生活。

1962年，鲍赞巴克在巴黎美术学院读书，受教于颇具戏剧色彩的鲍顿因开设的工作室。凭借自己的直觉，鲍赞巴克决定去参观汉斯·夏隆设计的柏林爱乐乐团音乐厅，这对他的整个职业生涯产生了显著影响。他（开始）尽量避免直角，并热衷于曲线——一切能够想像的曲线。

1964年，一部分人离开了鲍顿因的工作室，转投坎迪利斯门下。鲍赞巴克直到第二年才加入他们的行列，这是他与安托万·格伦巴赫弥笃友谊的结果。

尽管如此，新工作室的古典主义风格仍然让人难以忍受。鲍赞巴克感觉自己亟需休整，因此于1966年启程前往纽约，名为探究经典建筑，实则被那里的音乐

Without fail, however, they spent their holidays at the family residence outside Combourg, in Chateaubriand territory near St-Malo. This 16th-century manor with its 18th-century modifications was doubtless a significant factor in the shaping of the architect's vision, as were the nearby standing stones, prehistoric burial mound, wood and pond. On his father's return to civilian life, the family moved to Rennes and a new life began for Christian.

1962 found him in Paris at the Ecole des Beaux-Arts, in the atelier run by the highly theatrical Eugène Beaudoin. Following his intuition, Portzamparc decided to go take a look at Hans Scharoun's Philharmonic Theatre in Berlin, whose impact on him was to mark his entire career. Abandoning right angles, he gave in to the curve — to all imaginable curves.

In 1964 a splinter group left Baudoin's atelier for that of Candilis. Portzamparc did not join them until the following year, one outcome being his close friendship with Antoine Grumbach.

Nonetheless, the classicism of the ateliers was hard to put up with. Deciding he needed a break,

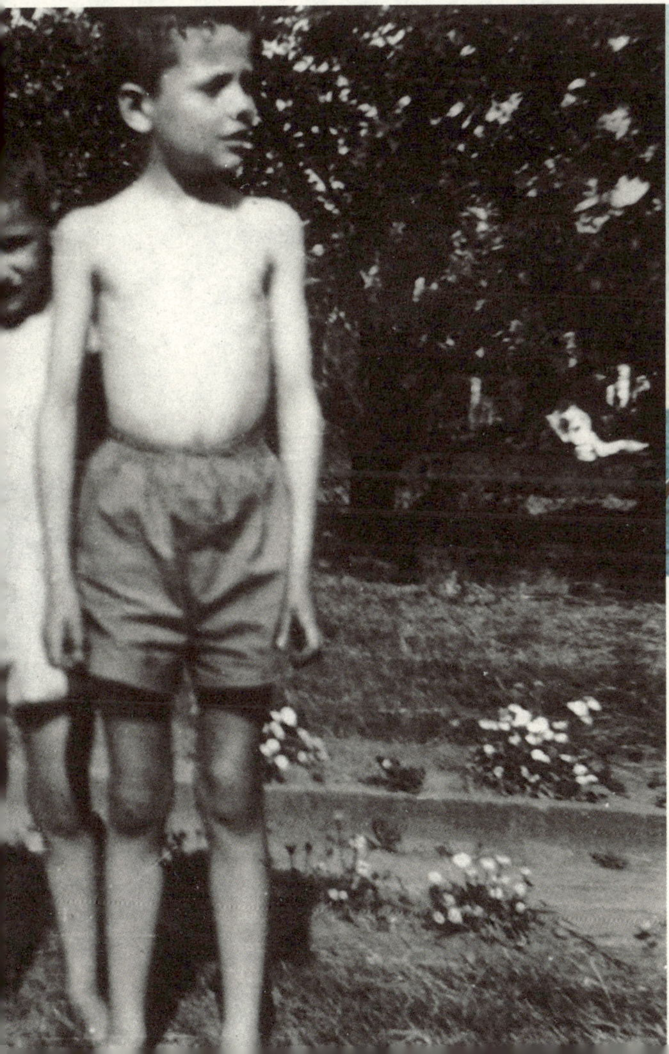

所吸引：克尔特林和蒙克，迪伦和地下丝绒乐队，卡索和金斯伯格，"垮掉的一代"和"沃赫尔工厂"。

他徘徊于穷乡僻壤之间，居住在纽约东部贫民区廉价的旅馆中，与即兴表演爱好者吉恩·雅克（Jean-Jacques）结为知己，投身反主流文化（运动），重拾他在巴黎时抛之脑后的政治立场，做酒吧服务员糊口……一整年的精神自由与随心所欲。

他回到法国，深信诗歌与绘画比建筑更为重要。1967 年的巴黎沸腾雀跃，什么事都有可能发生。他聆听一个又一个讲座——阿尔都塞，巴特，福柯，拉肯；参与国际情景画派；浸染各种意识形态和理论；也涉猎中国的文化大革命。回到艺术学院，他发觉身边拥有不少蓄势待发的艺术家和建筑师，吉恩-皮埃尔·布非，皮埃尔·毕赫葛里欧，罗兰·卡斯特罗，热拉尔·弗罗芒热，梅利·若利维，弗朗索瓦·雷昂，盖伊·德·卢日蒙；他们一起在职业委员会任职，因此，像当时很多其他人一样，没有参加考试就获得了学位——但并非没有吸收学院的学术思想。1968 年，鲍赞巴克放弃了绘画和素描而成为狂热的（文献）阅读者；这一时期（各类）讨论团体如日中天，旧的党派系统正在被独断的、对学说的忠诚所取代。鲍赞巴克也概莫能外，他与让-皮埃尔·布非，罗兰·卡斯特罗，安托万·格伦巴赫，盖伊·耐佐特，哲学家让-保罗·杜磊，工程师吉勒斯·奥利弗一同组成七人小组，集体致力于"将理论探索塑造为实践的基础与先导"的工作中。

无论如何，天下无不散之筵席，鲍赞巴克（最终）决定采纳毛泽东思想的基本原理作为逻辑性结论，全身心地投入"具体且实在的实验"中。多年来的"知识酝

鲍赞巴克，塞吉，菲利普。
Christian with Serge and Philippe.

塞吉，伊丽莎白，鲍赞巴克。
Serge, Elizabeth and Christian

he left for New York in 1966, officially to explore its architectural riches but in reality irresistibly drawn by the music scene: Coltrane and Monk, Dylan and the Velvet Underground, Corso and Ginsberg, the Beat Generation and Warhol's Factory.

He hung out in the Village, lived in the sleazy hotels of the Lower East Side, struck up a friendship with happenings-freak Jean-Jacques Lebel, plunged into the counterculture, found the political stance that had eluded him in Paris, made his living as a barman... A year of total intellectual freedom, of doing as he liked.

He came back with the conviction that poetry and painting were more important than architecture. In the effervescent Paris of 1967, everything seemed possible. He went from one lecture to another — Althusser, Barthes, Foucault, Lacan; got involved with the International Situationists; soaked up ideology and theory; and opted for China's Cultural Revolution. Back at art school again, he found himself with Jean-Pierre Buffi, Pierre Buraglio, Roland Castro, Gérard Fromanger, Merri Jolivet, François Rouan, Guy de Rougemont and others destined for fame as architects and artists; together they sat on the occupation committee and, like so many at the time, took out a degree without sitting the exams — but not without absorbing the school's values. 1968 saw him give up drawing and painting to become a feverish all-night reader; this was the golden age of the discussion group, with the old clan system being supplanted by dogmatic,

酿”纠正了他的形式主义倾向，但却无损于他对感觉（感官）的渴求。他总是将欲望置于节制之上。他的第一座建筑——位于马恩拉瓦莱的“绿色之塔”——就是明证。在建筑的苍穹中陡然升起一颗新星。欧风路（住宅）项目落成两年后，即1981年的一次聚会上，他巧遇一位聪颖的社会学家，她正跟随亨利·雄巴德劳维攻读博士学位，并参与一个城镇规划工作室。不到一年光景，他们便结为夫妇；1983年，塞吉出生了，紧接着弟弟菲利普于1986年降生。在此期间，伊丽莎白（Elizabeth，鲍赞巴克之妻名——译者注）的兴趣逐渐从城市社会学转向室内设计。她在蓬皮杜中心附近开设了Mostra展廊，着手设计家具，也展出其他艺术家和建筑师的作品，并开设自己的事务所。

伊丽莎白是克里斯蒂安的妻子，但她也时常是鲍赞巴克的业务合作伙伴，在项目中担当室内设计——例如格拉斯法院，魁北克图书馆，柏林的法国驻德大使馆，雷恩的科技文化中心等项目。

鲍赞巴克是一位地道的“左岸”巴黎人，曾在瓦诺（Vaneau）生活，也在穆费塔尔街、阿莱西亚街、梅德西斯街、苏弗伦大道居住过。他的事务所——自称为工作室——先后设在Rue de l’Abbaye、Rue de Seine、Rue d’Alésia、Rue de l’Aude。对他来说，行动高于一切。“不要理论，只将一切现有之物编纂综合——艺术的、文学的、科技的、情感的——以求在项目中博采众家之长。剔除理论，却敏于思考与反思。”他做出了抉择：不加入任何派系、小圈子、小集团。虽然这将招致连珠炮似的批评，以至伤害他或引起别人的误解，但他绝不卑躬屈膝。

doctrinal loyalties. Portzamparc was no exception to the rule, finding himself — with Jean-Pierre Buffi, Roland Castro, Antoine Grumbach, Guy Naizot, philosopher Jean-Paul Dollé and engineer Gilles Olive — in the Group of 7, collectively bent on "making theoretical exploration the basis of their practice".

All good things must come to an end, however, and Portzamparc decided to take the Maoist rationale to its logical conclusion in an utter dedication to "concrete experimentation". All these years of intellectual ferment cured his formalist leanings, but never killed his yen for the sensory. He would always set desire above moderation.

Proof came with his very first building, the "Tour Verte" water tower at Marne-la-Vallée. Out of the blue there was a new star in the architectural firmament. At a party in 1981, two years after the thunderclap of the Hautes Formes, he met a brilliant young sociologist who had taken her doctorate under Henri Chombard de Lauwe and was running a town planning workshop. Barely a year later they were married; baby Serge made his appearance in 1983 and brother Philippe came along in 1986. In the meantime, Elizabeth's interests gradually shifted from urban socilogy to interior design. She opened the Mostra gallery beside the Centre Pompidou, started designing her own furniture as well as bringing out other pieces created by artists ans architects and launded her own agency.

Elizabeth might have been Christian's wife, but she was frequently Portzamparc's associate as well, taking care of the interiors for him on such projects as the Law Courts in Grasse, the Library in Quebec, the French Embassy in Berlin and the multicultural center in Rennes.

Portzamparc is a thoroughly left-bank Parisian, having lived at Vaneau, and on the Rue Mouffetard, Rue d'Alésia, Rue de Medicis and Avenue de Suffren. His agencies — he calls them workshops — were on the Rue de l'Abbaye, Rue de Seine, Rue d'Alésia and, still, the Rue de l'Aude.

For Portzamparc, then, action was the thing. "No theorizing, but an interweaving of all the registers — artistic, literary, scientific, emotional — so as to get the most out of them in our projects. No theorizing, but endless thinking and reflection." And his mind was made up: he would be part of no clan, no clique, no coterie. This would earn him barrages of criticism that often left him hurt, wounded and misunderstood, but always unbowed.

As early as the La Roquette competition of 1974 Bernard Huet, while praising Portzamparc,

克里斯蒂安·德·鲍赞巴克与菲利浦·索莱尔。

Christian de Portzamparc with Philippe Sollers.

鲍赞巴克在大街的公寓。

The architect's Avenue de Suffren apartment.

早在1974年的拉·若凯特竞赛时，伯纳德·于厄在表扬鲍赞巴克的同时还评论道："他的方案没有勇气面对现实和周围的城市。在克里斯蒂安的宣言中有值得怀疑和容易引起遗忘的地方：他试图说服我们他实际上在创造新的空间类型，而非简单地摹写。他说要探求一种新的无时间特性的建筑风格——这似乎举足轻重。而对于阿尔多·罗西来说，华丽的修辞只是意欲掩盖怀旧情怀的华而不实的努力，这将导致工作陷入孤单、封闭的处境，敌视历史，固步自封以至只能反复自引用，至多不过禁锢于与现代城市的对白。"支持与反对并存的古怪态度。然而无论如何，这一评价在长时间内被很多人认同，使得鲍赞巴克被冠以"站不住脚的历史主义者"之名。

六年后的1980年，建筑师保罗·波特盖希在威尼斯组织了一次名为"新趋势"的聚会。在国际建筑届中稍具名气的人物悉数到场，他们中有弗兰克·盖里，迈克尔·格雷夫斯，汉斯·霍莱茵，矶崎新。

鲍赞巴克没有参加，但第二年的会事在巴黎的萨伯特教堂举行，成为"秋季欢庆节"的一部分，并拥有晦涩的题目——"活的历史"。这次，组织者米歇尔·盖伊以私人名义邀请鲍赞巴克参加，委托他设计展会入口。鲍赞巴克的设计类似某种谜语中的隐喻，与乔奇奥·德·基里科形而上的风格颇有几分神似。

在巴黎，反对的呼声烽烟再起，批评家香岱儿·贝蕾（Chantal Béret）谈及"老迈的现代者、年轻的古旧者，以及新的骑墙派（两面讨好者——译者注）之间的对峙。"

波特盖希拒绝邀请鲍赞巴克加入威尼斯后现代主义革新阵营的原因可见一斑：

commented, "The project doesn't dare to address the present and the surrounding city. There's something suspect and amnesic about Portzamparc's declarations: he's trying to convince us that he's actually producing new kinds of spaces, and not simply reproducing. He says he's looking for a new-style architecture that will be undatable — as if that was of any importance. As with Aldo Rossi, the rhetoric is just a fruitless attempt to cover up a nostalgia that turns the work into an aloof, closed universe, hostile to history, introverted to the point of self-quotation and at best closed to contemporary urban dialogue." An odd attitude, that: for and against at the same time. But whatever, the anathema had been pronounced and for a long time, and in the minds of many, the name Portzamparc remained sullied with the suspicion of an indefensible historicism.

Six years later, in 1980, architect Paolo Portoghesi organized a get-together in Venice titled "La Strada Novissima". Everybody who was anybody on the international architecture scene was invited, among them Frank Gehry, Michael Graves, Hans Hollein and Arata Isozaki.

Portzamparc did not take part, but the following year's event took place in the Salpêtrière chapel in Paris, as part of the Autumn Festival and with the ambiguous title of "Présence de l'histoire" ("Living History"). This time organizer Michel Guy made Portzamparc his personal guest, going so far as to entrust him with the design of the entrance. Portzamparc's contribution was a kind of metaphor of the riddle, somewhat in the metaphysical tone of Giorgio di Chirico.

In Paris the balloon went up again, with critic Chantal Béret speaking of a confrontation between "Old Moderns, Young Ancients and New Fence-sitters".

Portoghesi's refusal to conscript Portzamparc for his postmodernizing Venetian crusade was readily explainable: the Frenchman was challenging modernist orthodoxy, but there was nothing historicist about his approach. But getting himself invited along as Michel Guy's special guest brought the full fury of the

位于萨伯特教堂的装置，是"活的历史"展览的展品之一，1981 年。

Installation at the Salpêtrière, as part of the "Living History" exhibition, 1981

法国人（指鲍赞巴克——译者注）挑战现代主义者的信条，但他采取的途径无涉于历史主义。他成为米歇尔·盖伊的特邀嘉宾，这激起了反对者对他的愤慨。保罗·舍梅托夫将后现代主义等同于新保守主义的同时，还将鲍赞巴克作为明确的（攻击）对象。《建筑技术》学刊上暴风骤雨式的争论尚未偃旗息鼓，吉姆·帕莱特又掀起新的波澜。帕莱特在萨伯特教堂录制他的电视节目 Aux Arts Citoyens，真的将鲍赞巴克与舍梅托夫、让-路易·科恩、亨利·高丹、维多里奥·格里高蒂几人一同置于拳击台上。

很多年以后，在 1997 年，一封匿名（使用笔名）信函引起新一轮争论，信中对柏林的法国驻德大使馆设计竞赛的评选结果提出质疑，竞赛胜出者正是鲍赞巴克。"活的历史"引发的激烈争论和对鲍赞巴克的其他批评之焦点——尤其是他的拉·若凯特和欧风路两个项目——在于鲍氏的方案破坏了"已规划的城市"及"周边建成区域"的思维理念和设计原则，并由此产生焦虑与担忧。当时，柏林的法国大使馆建造工作正在酝酿，其过程处于怯懦、不切题、暗箱式的操作模式中，这对特定的法国工作模式提出严峻挑战。波及的受害者还包括：雅克，他最终失去了一个胜出的竞赛项目；弗朗西斯，他被剥夺了位于凯布朗利的国际会议中心的设计权；克劳德·瓦斯克尼，新任区域委员会主席迫使他中止瓦尔-德-马恩的公寓办公楼项目工作；最为严重的是——让·努韦尔，在"大赛跑场"事件（Grard Stade affair）中，只能和鲍赞巴克一道忍受来自政府部门的恶劣态度。这一切的粗俗简直毫无意义。而鲍氏则只能再次逆来顺受。

拳击台上：吉姆·帕莱特的 Aux Arts Citoyens 节目，电视秀，1981 年。

In the ring: Jim Palette's Aux Arts Citoyens TV show, 1981.

opposition down on him. In his equating of postmodernism with neoconservatism, Paul Chemetov took Portzamparc as an explicit target. A stormy debate organized by *Techniques et Architecture Magazine* was followed by another, just as tumultuous, orchestrated by Jim Palette. Palette recorded his TV program *Aux arts citoyens* at the Salpêtrière, literally putting Portzamparc in a boxing ring with Chemetov, Jean-Louis Cohen, Henri Gaudin and Vittorio Gregotti.

Years later, in 1997, an anonymous (because pseudonymous) circular did the rounds, calling into question the judges' decision in the competition for the French embassy in Berlin, won by Portzamparc. The real issue in the heated debate about "Présence de l'Histoire" and the criticism of Portzamparc — especially his La Roquette and Hautes Formes projects — was the nervousness generated by the planned city/built perimeter ideology. In the case of Berlin, what was afoot, in a particularly cowardly, irrelevant and underhanded way, was a challenge to certain French ways of working. Among the victims would be Jacques Hondelatte, who ended up losing a competition he had won; Francis Soler, dispossessed of the International Convention Center on the Quai Branly; Claude Vasconi, forced to stop work on the Val-de-Marne département offices by the new president of the Regional Council; and above all — above all — Jean Nouvel in the Grand Stade affair, with Portzamparc being made to bear the burden of the Government's bad manners towards Nouvel. The inelegance of it all was equaled only by its pointlessness. And once again Portzamparc had to grin and bear it.

This is not to say, however, that his career has been an endless succession of blows below the belt. As early as age ten, for example, he had taken out first prize in a sandcastle competition on a beach in Spain.

Since then prizes, distinctions, medals and other honors have piled up, among them two *Equerres d'Argent* (in 1988 for the Paris Opera Dance School at Nanterre, and in 1995 for the Cité de

杰伊·普利茨克为鲍赞巴克颁奖，法国文化部长雅克·特布出席。

Jay Pritzker presenting his prize to Christian de Portzamparc in the presence of Jacques Toubon, Minister of Culture.

与文化部长杰克·朗交谈。

A private word with Jack Lang, Minister of Culture

但这并不是说鲍氏的职业生涯总是无休止、连续地被“暗箭”所伤。在刚满十岁时，他就在西班牙某海滩的沙塔设计竞赛中夺魁。

自此以后，各类奖项、称号、奖章和其他荣誉如潮水般蜂拥而至，它们中包括：两次法国建筑银尺奖（1988年，楠泰尔巴黎歌剧舞蹈学校综合楼，The Paris Opera Dance School at Nanterre；1995年，则是“音乐城”和巴黎音乐舞蹈学校 Cité de la Musique and Dance Conservatory in Paris）；1990年获得法国巴黎建筑大奖；1993年荣获法国国家建筑大奖；最为著名的是1994年的普利茨克奖。随着国家建筑大奖的颁发，文森顿（Sybille Vincendon）在巴黎的《革命》(Libération) 日报上发表文章，褒扬“鲍赞巴克的加冕”和他20年来兢兢业业投入建筑事业的“情感与智慧”；而1994年5月16日的《时代》杂志上刊载了名为“反传统（主义）者的凯旋”的文章，宣称“鲍赞巴克异质性的、抒情诗般的作品获得了享有盛名的奖项”。

普利茨克奖自然非同小可，它是建筑界的诺贝尔奖。鲍赞巴克是首位，也是惟一一位折桂的法国人，同时还是最年轻的获奖者，他获奖时年仅50岁。评委会意见一致，一篇匿名新闻稿——似乎出自“高级女司铎”安达·路易斯·赫克斯泰勃尔的笔下——盛赞“一位造型诗人和深邃的空间塑造者的隽永才智，这智慧并非源于古典主义或现代主义，而只存在于独一无二的头脑当中。”让·努韦尔——与他自己的普利茨克奖近在咫尺——承认：“在我看来，鲍赞巴克的建筑是出类拔萃的电影导演的作品。一位最可信任的，能够胜任多种不同情境的导演，他使用严格从属于个人的‘武器’，例如破碎和分裂。”(Globe

la Musique's Music and Dance Conservatory in Paris); the City of Paris Grand Prix d'Architecture in 1990; the Grand Prix National d'Architecture in 1993; and most notably the Pritzker, in 1994.

The awarding of the Grand Prix National led Sybille Vincendon, writing in the Paris daily *Libération*, to speak of the "coronation of Portzamparc" and of the "emotion and intelligence" he had been bringing to building for 20 years; while in *Time* magazine of 16 May 1994, Margot Hornblower titled her Pritzker article "Triumph of an Iconoclast", following on with "Portzamparc's idiosyncratic and lyrical work is rewarded with a prestigious prize."

The Pritzker Architecture Prize, of course, is not small beer, being the architectural equivalent of the Nobel. The first and, so far, only Frenchman to have won it, Christian de Portzamparc was also, at 50, its youngest laureate. The jury's decision was unanimous and an anonymous press release — which nonetheless betrayed the hand of "high priestess" Ada Louise Huxtable — praised "the powerful talent of a poet of form and a creator of eloquent spaces, rooted in neither classicism nor modernism, but in a singular mind". Jean Nouvel, himself not far from taking out his own Pritzker, avowed that "In my opinion Portzamparc's architecture is the work of an auteur. One of the most authentic of all auteurs and someone who has been able to work in different situations using weapons, such as fragmentation, that are strictly his own." (*Globe Hebdo*, 1 June 1994)

It was in Columbus, Ohio, that Mecca of modern and contemporary architecture, that the supreme distinction was conferred on him by Jay Pritzker. Among those present was Jacques Toubon, then France's Minister of Culture, whose lightning trip testified to his friendship with Christian de Portzamparc and his lively interest in architecture.

Back in Paris, Elizabeth organized a big, secret party at the *Café Beaubourg*, attended by all the old friends from here, there and everywhere.

The perimeter of the Parc Montsouris in Paris is a fascinating precis of the private-sector architecture of the 1930s. Along its flower-

与巴西建筑师奥斯卡·尼迈耶合影。
With Brazilian architect Oscar Niemeyer

与日本建筑师安藤忠雄合影。
With Japanese architect Tadao Ando

Hebdo，1994 年 6 月 1 日）颁奖典礼在俄亥俄州哥伦布市举行，鲍赞巴克在这个现代建筑和当代建筑的麦加圣地被杰伊·普利茨克授予至高的荣誉。出席嘉宾包括时任法国文化部长的雅克·特布，他的莅临显示了他与鲍赞巴克之间的友谊和对建筑学的浓厚兴趣。

回到巴黎，伊丽莎白在波布尔咖啡厅组织了一次大型私人聚会，散落各处的同道旧友纷至沓来。

巴黎的蒙苏里公园一带是 1930 年代私人建筑（设计）部门的引人入胜的缩影。

福冈会晤。从左至右：马克·麦克斯，雷姆·库哈斯，矶崎新，史蒂芬·霍尔，克里斯蒂安·德·鲍赞巴克，榎本和彦。

Meeting at the Fukuoka summit. From left to right: Mark Macks, Rem Koolhas, Arata Isozaki, Steven Holl, Christian de Portzamparc and kazuhiko Enomoto.

lined streets and shady lanes are scattered the elements of a kind of modernist manifesto.
Overlooking the Avenue René Coty, as if clinging to a rocky outcrop, is a strange white house, all overhangs and breaks in the facade: very "esprit nouveau" in fact, and dating from exactly 1930. It is the work of the Breton painter Le Mordrant, whose seascapes and naturalist fishing scenes make a visit to the Quimper Museum worthwhile.
Le Mordrant had become blind after being gassed in the First World War and legend has it that, modifying one model after another, he designed his house by touch. Much later a

位于 Rue de l'Aude 的工作室，在巴黎蒙苏里公园附近。

The Workshop on the Rue de l'Aude, near the Parc Montsouris, in Paris

在花团锦簇的街道旁，在树影婆娑的径路边，激荡着现代主义者的宣言。

沿 René Coty 大道远眺，有一座造型奇特，好似附着在突兀出地面的巉石上的白色小屋，立面上尽是悬挑与凹陷：很具有“新新智慧”，而实际上它是 20 世纪 30 年代遗物。这座建筑出自布里多尼画家勒·莫丹特之手，这位画家的海景和自然主义捕鱼场景绘画是坎佩尔博物馆的收藏展品。

勒·莫丹特在第一次世界大战时中毒

near-miracle gave him back his sight. Enthused by the "events" of May 1968, and in spite of his advanced age, he took to the Rue Gay-Lussac one night — and could possibly have met a certain young architect there. That night the tear gas cost him his sight again.

Now another Breton, a skilled sailor and one hell of a helmsman, lives in the house, surrounded by a variable-geometry team currently peaking at fifty. Among them are the faithful, the unbudgeables, like François Barbereau, Bertrand Beau, Frédéric Binet, Carole Claverie, Benoît Durbek, Benoît Juret, Marie-Elisabeth Nicoleau and Etienne Pierres. Others, like albatrosses, have

气失明，传说他用双手一次次触摸并修改模型，设计了自己的住所。很久以后，他奇迹般地恢复了视力。后来，他受1968年“五月事件”的激励，不顾年事已高，在一天晚上去了盖-吕萨克街——很可能约见了一位建筑师。那晚，催泪瓦斯使他再次失明。

现在，另一位布里多尼人，一位技能卓绝的水手和舵手，居住在这所建筑里，周围团结着一个灵活多变的团队，如今规模已达50人，人丁鼎盛。在他们当中有一如既往的忠实追随者，像François Barbereau，Bertrand Beau，Frédéric Binet，Carole Claverie，Benoît Durbek，Benoît Juret，Marie-Elisabeth Nicoleau，Etienne Pierres。也有的人仿佛信天翁一般，展翅启程投奔异域：例如Frédéric Borel，François Chochon Julie Howard，Sam Mays ou Richard Scoffier三人。

这些合作者和友人朝夕相处，共同为设计工作出谋划策，带来自己独特的感受、视角和敏感性。你能发觉他们的贡献贯穿于（事务所）的整个发展历程：在巴黎的音乐城；在柏林、格拉斯和雷恩；在东京和纽约；在卢森堡、里尔、首尔、堪萨斯城、阿尔梅勒市、魁北克、蒙彼利埃、里昂——（城市）清单仍在延伸。

他们在未曾实现的项目中也同样崭露头角。例如威尼斯双年展的法国馆设计中，鲍赞巴克与时任波尔多现代艺术中心负责人的让·路易斯·佛罗门特共同努力，使鲍赞巴克、努韦尔，和菲利普·斯塔克三人组成了让人津津乐道的“铁三角”。还应提到在阿姆斯特丹独一无二的历史文脉中极尽现代气息的电影院：它并非纯粹是“光的试验”，而是对帕拉第奥精彩卓绝的颂词。此外，位于香榭丽舍的雪铁龙项目将莫比乌斯带（的形体）结合景深，创造

为威尼斯双年展设计的法国馆，1990年。
Project for the new French pavilion at the Venice Biennale, 1990

1990年威尼斯设计竞赛时与努韦尔，菲利普·斯塔克合影。
With Jean Nouvel and Philippe Starck at the time of the 1990 Venice competition.

taken wing for fresh voyages and other climes: Frédéric Borel, François Chochon Julie Howard, Sam Mays ou Richard Scoffier, to name but three.

In this mix of collaborators and friends, everyone brings his own touch, eye and sensibility to the grand design. You find their contributions throughout the career: at the Cité de la Musique in Paris; in Berlin, Grasse and Rennes; in Tokyo and New York; and in Luxembourg, Lille, Seoul, Kansas City, Almere, Quebec, Montpellier, Lyon — and the list goes on.

They are equally present in those projects that never got off the ground. The French Pavilion at the Venice Biennale, for instance, with Jean-Louis Froment, then in charge of the Contemporary Visual Arts Center in Bordeaux, organizing a stylistically delicious triangular between Jean Nouvel, Christian de Portzamparc and Philippe Starck. Then there was the highly contemporary cinema to be slipped into the historically unique context of Amsterdam; despite being a pure exercise in the use of light, the proposal comes across as a vibrant tribute to Palladio. And what about the Citroen project on the Champs Elysées, which brought a rare degree of fantasy

出绝无仅有的幻想效果和幽默情调。

这是三处十分另类的设计，叙写了众多古灵精怪的主题、测度、张力、维度、透明度、抽象与具象性、出现与消失等内容，还涉及技术与风格、图像与声音、感觉与情感的拓展——简而言之，构成一名建筑师日常创作环境的一切内容。

然而，正像朱尔斯·雷纳德所深知的那样，箴言有曰：思想本身空无一物！没有语言，它总是苍白无力的。

and humor to its combination of the Moebius strip and depth of field?

These were three very discursive proposals, full of queer stories of inscription, measure, tension, dimension, transparency, abstraction and concreteness, appearance and disappearance, effacement, and expansion of technique and style, images and sounds, sensation and emotion — in short, everything that goes to make up the architect's daily round.

But there are words, too, as Jules Renard knew well: "Ideas are so empty! Without the words, no interest at all."

阿姆斯特丹一座电影院方案。

Model of the Bandai tower, used by Cédric Klapisch in his film Peut-être (2000)

“万代”大厦模型，成为塞德里克·克拉皮什在电影《或许》(Peut-être, 2000) 中的配景。

Project for a luminous cinema in Amsterdam

在位于香榭丽舍的雪铁龙项目中设计的莫比乌斯带。

Moebius strip for the Citroen project on the Champs Elysées

2

实验
experimenting

位于拉维莱特的两个独立实体：西侧，国立音乐学校，东侧，音乐之城。

Two separate entities at La Villette: to the west, the National Conservatory of Music, to the east, the Cité de la Musique.

音乐城

The Cité de la musique

许多人将音乐学校和音乐城当作鲍赞巴克事业的顶点和登峰造极之作。实际情况并没有这么简单，我们还将进一步体悟到事情的复杂性：这里所有的至高成就严格来讲都是暂时性的，也只是细节问题而已。

试想一位年轻的钢琴家在获奖（Margaret Long Prize）时却只是乳臭未干的黄毛小子：1984 年赢得音乐城竞赛时，鲍赞巴克刚及不惑之年——以建筑学的标准衡量仍然是孩童而已，顶多相当于画家或音乐家的弱冠之年。

的确，他可以塑造（建筑）比例，表

Many regard the Conservatory and the Cité de la Musique as a crowning achievement and the high point of Portzamparc's career. Things are not quite as simple as that and further on we shall see just how complex they are: any high point to be found here was strictly temporary and a matter of detail.

Try to imagine a very young pianist taking out the Margaret Long prize and becoming a case of arrested development the very same day: for when he won the Cité de la Musique competition in 1984, Portzamparc was barely forty — still a child in architectural terms and the equivalent of a musician or painter at age 20.

True, he could play his scales, display an individual temperament, demonstrate sensitivity

拉维莱特的牲畜交易市场。
The cattle market at La Villette.

两座建筑组成了拉维莱特公园引人注目的入口空间。
The two sections make up a striking entrance to the La Villette park.

达个人气质，展示敏感性和表现力，证明自己拥有必要的手段和方法；然而，他的事业是蒸蒸日上还是江郎才尽呢？真正的事业难道能离开不知疲倦地探索么——排练、音乐会、经历和旅行——（正是这种种）探索带来多样性，（并使人）推陈出新和发展演变。

尽管如此，考虑到其规模、全面性和复杂性，考虑到它享有的声誉，以及它敢于在拉维莱特这座"小城市"中自居为"城"的事实——并且有幸在建成之时便随即得到普利茨克奖的青睐——"音乐城"被广泛公认为（鲍氏的）代表作也不足为奇。

是登峰造极之作也好，是阶段性成果也罢，肯定存在某个开端，这一开端时刻肇始于拉维莱特地区从统一走向分化之前。

1867 年——更确切地说，10 月 22 日，拿破仑三世、莫尼公爵和作家普罗斯佩·梅里美出席——位于潘廷门的牲畜交易市场和位于拉维莱特门的屠宰场举办了官方开市仪式。这里位于巴黎东北地区，城市与农村迎面相击，就像闭塞的、使用蹩脚方言的农夫行走于已经工业化的世界一般。这是两个领域、两种心智的汇合。

and expressiveness and prove he had the necessary technique; but was his career ahead of or behind him? Doesn't a career call for the ongoing discovery — the rehearsals, concerts, experience and travel — that will bring diversity, innovation and evolution?

Even so, in its size, its fullness and its complexity, in its prestigiousness, in its title as a "City" within the "little city" of La Villette — and with the Pritzker coinciding with its completion — it is unsurprising that the Cité de la Musique was widely regarded as a culmination.

Whether culmination or simple staging post, there had to have been a beginning somewhere. And that beginning lay in the time before the time when La Villette's single territory became many.

The year 1867 — on 22 October to be precise and in the presence of Emperor Napoleon III, the Duc de Morny and writer Prosper Mérimée — saw the official opening of the cattle market at Porte de Pantin and the abattoirs at Porte de la Villette. Here in northeastern Paris country and city met head—on as be-clogged, patois-speaking peasants rubbed shoulders with a world already turned industrial. This was the confluence of two different realms, two mentalities. Over at the Porte de Pantin cattle market, where the Conservatory and the Cité de la Musique stand today, the bell rang for the close of business at one o'clock exactly, and the men in blue

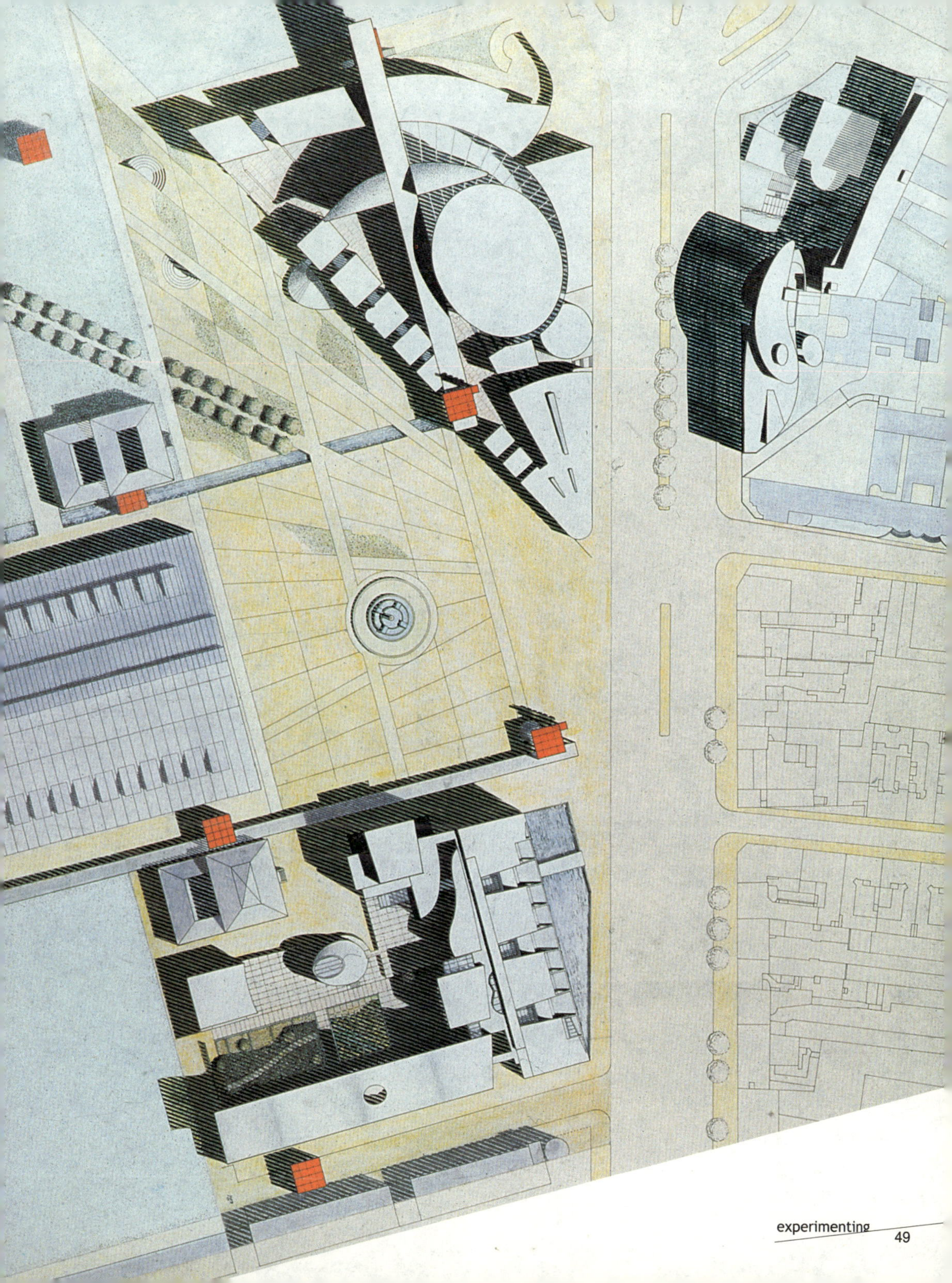

音乐学校洁白绵延的立面：可以穿越的城墙壁垒。
The long, white curve of the Conservatory facade: a rampart that can be crossed.

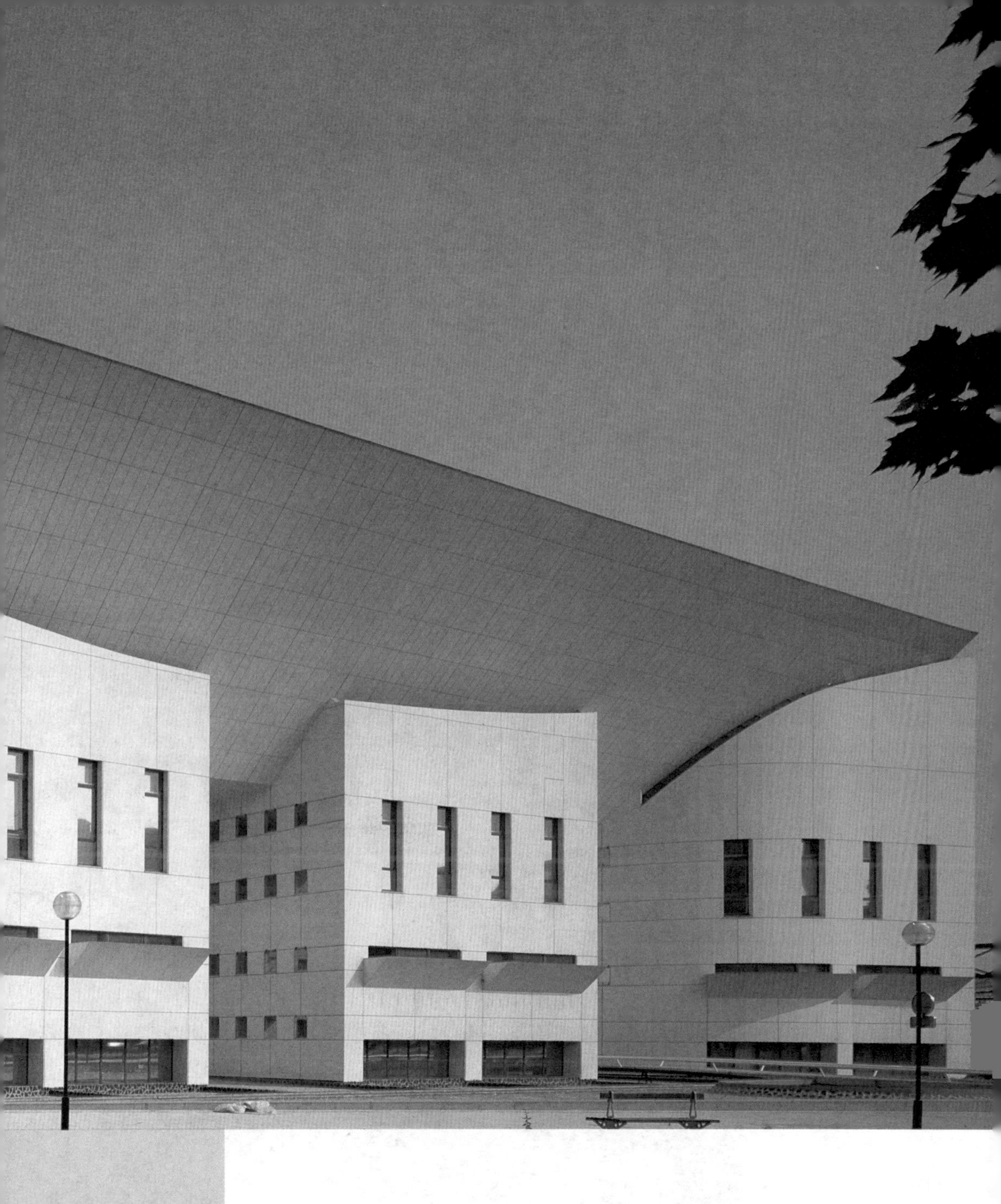

鲍赞巴克的得意之作：倒映于水中的立面。

A Portzamparc favorite: the facade reflected in water.

透明：光线塑造一切，一切又映衬光线。
Openings and transparency: here everything exists through and for light.

在潘廷门的牲畜交易市场——也就是今天音乐学校和音乐城坐落的地方——闭市的钟声准时在一点钟响起，身穿蓝色工服、佩戴黑色高帽的工人放下手中的铁钩，锥刺和用于标记所买牲畜的包盖着皮革的剪子。他们沐浴更衣，擦亮鞋子，赶集一般涌入大街（Rue d'Allemagne）上的餐厅，即现在的大街 Avenue Jean Jaurès：the Cochon d'Or，the Dagorno、the Horloge、the Bœuf Couronné。这被戏称为"随钟鸣而作息"。

这里你会与饲者、马贩子、牧人、驯

smocks and high-crowned black caps put away their crooks, their prods and the leather-covered scissors used for marking the animals they had bought. They changed clothes, got their shoes shined as if going on the town and crammed themselves into the restaurants on the Rue d'Allemagne, now Avenue Jean Jaurès: the Cochon d'Or, the Dagorno, the Horloge and the Bœuf Couronné. "Going with the bell", it was called.

Here you met breeders, horse-copers, graziers, handlers, bankers, money-lenders: confident, competent people, proud of having a trade you couldn't learn in any school and radiating an aura of everlasting power.

音乐学校的风琴演奏厅。
The Conservatory organ hall.

鲍氏的构思草图已然精准细腻。
Already Portzamparc's preparatory drawings are strikingly precise.

兽师、银行家，放贷者不期而遇：他们都是满怀自信，精明能干的人，从事着在任何学院中都无法教授的贸易活动，散发着亘古不灭的活力。

然而，事事变迁无可避，尤其是城市，其本身的特性便是膨胀、扩充、伸展、充溢。

1977 年交易市场和屠宰场停业了，它们迁移到城市边界以外的新场地。遗留下来的占地 55 公顷的拉维莱特地区徐徐改变身形，而乌尔克运河是交易市场和屠宰场之间的边界——生与死的疆界——此后成为拉维莱特地区的主要交通线。

随着市场和屠宰场的停业，一项科学博物馆计划诞生了，准备利用拉维莱特门较新但不甚完善的屠宰场建筑。又有传说称这里将建设一座公园。然而 1981 年风云突变，弗朗索瓦·密特朗出任总统，杰克·朗担任文化部长：一切正在筹划的文

Ah, but nothing is unchangeable; especially a city, which by its very nature swells, expands, extends, overflows.

In 1977 the market and the abattoir closed, moving to fresh fields beyond the city limits. Gradual change came to the 55 hectares of the territory of La Villette, and the Ourcq Canal, frontier between the market and the abattoir — between life and death — became the main thoroughfare of what would henceforth be called the "territories" of La Villette.

Out of the closure was born a plan for a Science Museum that would use the new, if incomplete abattoir building near Porte de La Villette. There was vague talk of a park to round things off. Then all of a sudden it was 1981, François Mitterrand was President and Jack Lang was Minister of Culture: all current cultural projects were put on hold, re-assessed and, in line with the subsequent recommendations, cancelled, confirmed or expanded.

At La Villette, expansion carried the day. In a few

化性项目被暂时搁置，重新评价，接下来根据相应的建议和结论，项目被取消、批准或扩展。

在拉维莱特，扩建与拓展的呼声如日中天，成为主旋律。短短几个月时间，规划中的科学博物馆变为科学技术与工业博物馆，增设360度环幕电影院，以昂其身价。在运河南侧地区计划营建有顶大展厅和珍妮特（Zénith）娱乐综合体，当然还有最值得关注的音乐城。这些项目将统一安置在一座标新立异的城市公园内，公园内遍布“奇珍异宝”和各类主题园，兼备动感跳跃和敛心默祈式的气息与氛围。

设计竞赛如期举行，参与并胜出者包括：范新伯赢得博物馆和环幕电影院的设计权。Reichen和罗伯特负责大展厅，Chaix和Morel则设计珍妮特娱乐综合体（顶点中心），屈米（Tschumi）负责公园的规划。

short months the planned Science Museum became the Museum of Science, Technology and Industry, with the Géode and its 360° cinema thrown in for good measure. For the area south of the canal the ideas were the Grande Halle covered market, the Zénith entertainment complex and, especially, the Cité de la Musique. The setting for all of this would be a brand new kind of urban park, scattered with "curiosities" and theme gardens in a mix of the busy and the contemplative.

The competitions were organized, competed in and won: Fainsilber for the museum and the Géode, Reichen and Robert for the Grande Halle, Chaix and Morel for the Zénith and Tschumi for the park.

And in 1984, Christian de Portzamparc was chosen for the Conservatory and the Cité de la Musique: a colossal project whose unparalleled richness and complexity would signal a vital phase in the young architect's career.

Firstly because he realized that the two-stage

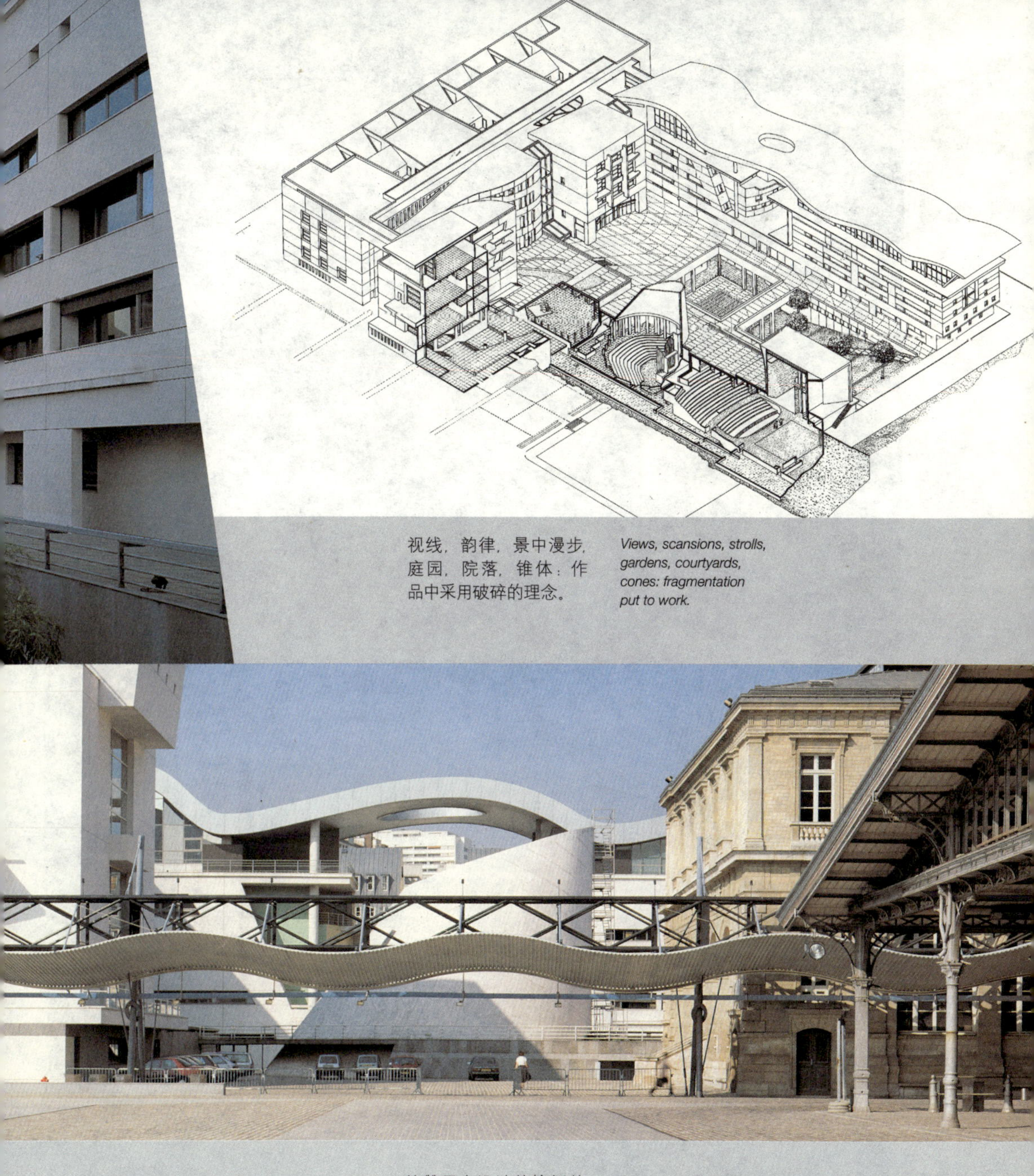

视线，韵律，景中漫步，庭园，院落，锥体：作品中采用破碎的理念。

Views, scansions, strolls, gardens, courtyards, cones: fragmentation put to work.

鲍赞巴克设计的蜿蜒的波形（屋面），与屈米（构思的）贯通整个公园的步行区域相得益彰。

Portzamparc's long wave is matched by Bernard Tschumi's pedestrian precinct, crossing the park from one end to the other.

1984 年，鲍赞巴克被委托设计音乐学校和音乐城：这项鸿篇巨制一般，空前辉煌和错综复杂的设计，诠释了年轻建筑师职业生涯的重要阶段。

首先，建筑师意识到这项分两期进行的设计将占据他生命中的 10 个年头。其次，目前的 7 人工作团队必须在一夜间拓展至 25 人。

总而言之：（建筑师）的时间计划和团队结构产生了剧烈震荡；面对全新的课题；而且必须发展新的经营模式。与前述的楠泰尔舞蹈学校相比，局面是前所未有的——显而易见——尤其自巴士底歌剧院

completion process was going to take ten years of his life. And secondly because the current team of seven would have to expand, overnight, to twenty-five.

To sum up, then: a real upheaval in terms of timeframe and structure; entirely new issues to be faced, and new modes of functioning to be developed. Compared to the preceding experience with the Dance School at Nanterre, this was a whole new ballgame status—wise — especially since the loss of the Opéra Bastille competition, which had left Portzamparc and his team in a state of advanced fatalism, with no great hopes for La Villette. It may well have been, of course, that this spirit of detachment brought a relaxedness, an ease and a lightness of touch

整座音乐学校充满了室内与室外的形式辩证。

A formal dialectic between inside and out is generated throughout the Conservatory.

恰似活的生物体一样，建筑表达了身体与眼睛，精神与肉体的主题。

Like a living organism, architecture addresses the body and the eye, the mental and the physical.

竞赛失利以来，鲍赞巴克及其团队陷入了深层的宿命论（心理）状态，他们对拉维莱特并未抱有必胜之信心。当然，很可能是这种超然心态营造了无拘无束的心境，一份从容，一份举重若轻的感动，最终迎得了评委的赞许。下一步就是处理冗繁缭乱的设计工作了。欧风路和楠泰尔舞蹈学校已然功成，鲍赞巴克实践了“开放街区”的理念，或被普遍称为“白色建筑”的设计思想；如今良机再现，机缘巧合，他自然要更上一层楼。

整整十年以后，1994 年举行的音乐城落成典礼上，鲍赞巴克强调“人类是整体的生命：耳不离目。贯穿这一项目的十年间，我常常思考听觉与视觉的关系，思考这座建筑使用者身体的、视觉的、和精神的活动”。

“有人业已向我评论说音乐学校拥有 40000 平方米的私密（空间）。这从很大程度上得益于（建筑）细分为多个单元——建筑中的建筑——每一单元是独立的声学体。此外，我对供学生交往的开放交通空间进行的总体性设计创造，也加深了这一印象。这些理念在音乐城的新建部分中得以发扬和拓展，它即将向公众开放。从‘活的有机体’的概念发端，我从一开始就制定了基本原则，这一部分的音乐城将允许局部加建、扩展或缩减。您可以将这个项目的酝酿和构思比拟一座港口，驳船是音乐演奏区域，而水体则是开放的交通空间。”

“我的第一要务是塑造满足不同听众多种要求的音乐空间。真正的挑战是依靠建筑自身来触发“音乐城”使用者的感官知觉，使人们徜徉其间，陶冶性情；我希望人们在这里可以感受到音乐的多元特征。有些建筑，你只需审视其部分空间，就会自觉身处一架巨硕的机器之中，很容

that ultimately tipped the scales for the judges. The next step was to come to terms with the dazzling scale of the thing. The Hautes Formes and the Nanterre Dance School were already there and Portzamparc had experimented with his "open block" and what's generally called "white architecture"; and all of a sudden there was the chance to go even further.

At the opening of the Cité de la Musique in 1994, a full decade later, Portzamparc affirmed that "Man is a total being: the ear needs the eye. Throughout the ten years of my life this project has taken up, I've thought constantly about the relationship between hearing and vision, about the bodily, visual and mental movements of the people who would be using the Cité de la Musique.

"Someone has commented to me that the Conservatory is 40,000 square meters of intimacy. This impression is largely the result of the subdivision into elements — buildings within the building — of which each is an acoustic envelope, and the resultant creation of open traffic areas for students to meet in. The idea is developed and broadened in the new part of the Cité, now open to the public. Working from the idea of a living organism, I envisaged right from the start, for this part of the Cité, a basic principle allowing for evolution, for expansion or reduction of some of its parts. You could describe the gestation of the project as a harbor, with the hulls of the ships as the music areas and the water as the open traffic spaces.

"What I had to do was come up with music spaces to suit every kind of audience. The challenge was to use the architecture itself to trigger the sensory perception of the people inside the Cité, get them moving, change their way of being; and I wanted music to be felt there as something pluralistic. There are buildings you only need to see a part of to have the feeling of being inside an enormous machine with a perfectly predictable overall plan. That was what I set out to avoid: I wanted people to discover the place, move through it. The Cité comes across as a sort of dream-village: there are streets and

（与实体）相辅相成的开口与镂空具有精湛的整体性气质。
Combining openings in a spirit of total virtuosity.

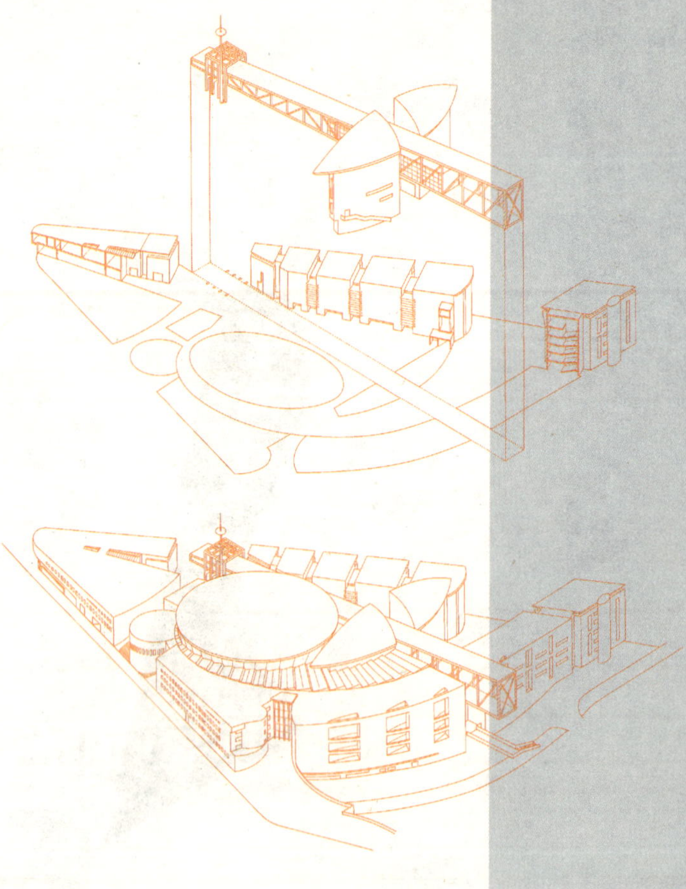

音乐城的多个体量被一座街道式的展廊统帅起来，庇护在一副通长的金属梁架之下。

The Cité de la Musique's multiple volumes are sunified by a street-gallery sheltering under a long metal beam.

易通晓其全盘。而这是我竭力避免的：我希望人们发掘建筑中的场所，往返遨游。'音乐城'意在塑造某种梦的城市：有街道、广场，有不透明的区域，也有沐浴着阳光的开放部分。它好似设有数个开口的贝壳；在壳体中，一座巨大的螺旋形展廊引领人们进入被清晰限定的、封闭的、各具其用的独立建筑中。这些封闭的空间包括音乐厅、音乐博物馆及其宣讲厅、当代音乐合奏团的办公室（offices of Ensemble Intercontemporain，也称现代乐集——译者注）、管理区、化妆间和排练区、学生宿舍、音乐咖啡厅和展览空间。"

"对公众开放的区域拥有多个出入口，它的中心是一座传统形式的中庭：我拓展这一构思，设计了围绕音乐厅的螺旋形空间。我希望这一区域在声学上出类拔萃，真正吸引人们竖起耳朵洗耳恭听：这也是我尝试牛角形的原因，它易于传递声音。从音乐维度讲，这座建筑是完整的组曲，由一系列有待使用者发掘的场所和旅程构成。与音乐中的情况类似，建筑在时间序

squares, opaque built sections and open areas under skylights. The Cité is built like a conch, with openings at several points; and inside the shell an enormous spiral gallery leads to the clearly defined, closed buildings, each of which has its own function. The closed spaces are the concert hall, the music museum and its lecture theatre, the admin area, the offices of the Ensemble Intercontemporain, the dressing rooms and rehearsal areas, the student accommodation, the music café and the exhibition spaces."

"The parts open to the public, have numerous entries and at their center is the idea of a traditional lobby: I've expanded this idea via the spiral that winds around the concert hall. I wanted this area to be acoustically distinctive so that people would really prick up their ears: that's why I tried this amplifying-horn shape, to transform the sound. The Cité as a whole is a suite in the musical sense, a succession of itineraries, of places to be discovered as you move on through. In architecture as in music, things are perceived in temporal sequences, in durations."

"I just love the fact that the architecture here gives you the urge to move about: you go from one place to another because you want to discover things. Traditional concert halls often bother me. people are pinned down like

幽深的连廊灯火通明。
At nightfall the long passageway lights up.

列中被持续性地认知。我很欣喜这座建筑能够激励你徘徊其间：你从这里行走到那里，试图做出新的发现。传统的音乐厅常常使我心生厌倦：人们像被擒的蝴蝶一样动弹不得。但在这里你可以环顾四周，目力增强、移步换景。我希冀这座建筑慷慨绝伦，激发各种感知——甚至肉欲的感知。因为我们的感知范围愈广泛，生活也就愈美好。"

音乐学校/音乐城的组合落成10年以后，我们有理由承认它们达到了预期目标，完成了既定任务，无负于大众的期待。

它们由两部分相互独立的实体组成：西侧是音乐学校，隐蔽于市井之间；东侧是音乐城，广迎八方来客。

音乐学校蜿蜒的立面修长而洁白，时

butterflies. Here the eye travels, the gaze dilates, shifts. I wanted the architecture to be generous, stimulating for the senses - sensual, even. The broader our range of sensory input, the better we live."

Ten years after delivery of the Conservatory/Cité ensemble, we can only acknowledge that its goals have been attained, its mission accomplished and its hopes rewarded.

Two separate entities, then: to the west the Conservatory, closed to the public; and to the east the Cité, open to all.

The incurvated facade of the Conservatory-long, white, punctuated with transparent areas and handsomely reflected in its moat-achieves a simultaneous intensity of lightness and density. Yet it speaks out frankly, "reads" easily: this host of openings and apertures is an exercise in trompe-l'oeil, for these are the loopholes of a contemporary fortress. The

中庭的构思发展为围绕音乐厅的螺旋形空间（设计）。
The idea of a lobby is expanded into a spiral surrounding the concert hall.

时为透明区域所打断，款款倒映于水池之中——兼备轻盈与密集的特征。它坦诚自若地宣称：这些群集的开口和空穴是错视画派的习作，当代堡垒的瞭望孔。这座建筑并不讳言自身特立独行的气质：在亲切热忱、却又兼具稳重婉拒意蕴的立面背后，2000 余人生活和工作其中，分享着教室、排练厅、体育馆、媒体中心、咖啡店、宿舍、会议室和幽闭的私人空间。

在立面的另一侧，四个独立的南北向单元容纳了音乐学校的各项功能。它们光鲜明亮，“相互间联系便捷”，使用卓越的手法为开口、视线、庭院、室内方庭所分解，采用文雅端庄的方式处理不透明性与光的主题。

最出彩的部分——圆锥形的风琴演奏大厅——破土而出，拔地而起，而西侧绵长奇特的波形屋顶将天穹半遮半掩。

音乐城则与此迥然不同。它貌似与音乐学校相互抵触，实则完善了整个构图，建筑由许多相互并置，变化多端的体量组成，其核心统帅是一座展廊/街道，被一副贯通整个音乐城的金属梁架所庇护。

音乐城向公园方向开敞，构图灵活自由，较之音乐学校缺少几分大家风范，却凭添几重从容与诙谐。

无论如何，两座建筑岿然对视，交相辉映。它们超越纯粹建筑维度的讨论，成功塑造了拉维拉特公园杰出的入口空间，映衬出“醒狮喷泉”和大展厅的立面。它们深深镌刻在场地之上，再造了“领域”观念，尽管这里不再有政治、经济、社会的云雨浮沉。

就在鲍赞巴克着手音乐城工作的时候，另外两个他自认为更为完满的项目也同时启动。在波布尔咖啡厅和布戴尔博物馆设计中，他沉湎于色彩与纹理的实验，抛弃“白色建筑”这一风行一时的特征。

从音乐角度理解，整个建筑构成了完整的组曲，由一系列有待使用者发掘的场所和旅程构成。

The entire Cité is a "suite" in the musical sense: a series of itineraries and places to be discovered as the visitor is drawn through it.

building makes no attempt to conceal its exclusive character: hidden behind a facade that is amiable and welcoming, but at the same time solidly resistant, more than 2000 people live and work, sharing classrooms, rehearsal rooms, a gymnasium, a media center, a cafeteria, accommodation, meeting points and private nooks.

On the other side of the facade four north-south bays accommodate the Conservatory's activities. Stunningly lit and "transit-friendly", they are skillfully broken up with openings, views, gardens and interior courtyards, bringing an elegant intelligence to the treatment of opacity and light.

The ultimate highlight — the conical organ hall — seems sprung directly out of the ground, while the long, strangely undulating roof of the west bay lets the sky in halfway along.

Things are different in the Cité. Seemingly running counter to the Conservatory, but in fact rounding off the composition, is the host of confronting, varied volumes whose unifying agent is a street/gallery, sheltered by a metal beam running the entire length of the Cité.

Opening onto the park and geometrically freer, the Cité may be less virtuoso than the Conservatory, but it is certainly more relaxed and witty.

In any case, in their combination of assonance and consonance, and above and beyond all purely architectural considerations, the two units make up a remarkable entrance to the La Villette park, perfectly setting off the lion fountain and the facade of the Grande Halle. Their stamp is on the site in their recreation of the notion of "territories" that political, economic and sociological ups and downs had banished.

As Christian de Portzamparc was settling down to work on the Cité de la Musique, two other projects got under way which he regards as more fully successful. With the Café Beaubourg and the Musée Bourdelle, he indulged in color and texture experiments far removed from the "white architecture" that had, perhaps, characterized him until then. The result was

其成果惊世骇俗，也偏离了音乐城的模式。十年之中——1984年至1994年——他的“主要工作”在潘廷门开展，但（我们）若将全部视野仅限于此则显得管窥蠡测了。他的想像力、感官、热情和探索绝不拘泥于此。

需要证明的话，鲍赞巴克的其他项目可以提供表达方式上显著的多元性/丰富性/多样性。

这些项目包括波布尔咖啡厅和布戴尔博物馆，也包括欧罗迪斯尼饭店，波尔多的月之港，日本福冈集合住宅（1989年），里昂纳信贷银行大厦（Crédit Lyonnais tower at Euralille，1992年）——不可穷举。

雄奇、精巧、宏伟庄严，音乐城的缔造者说道，机遇和技巧同样重要。广泛吸收各种流派与风格，音乐城在开始时难于被预料和控制。时光流逝，这座建筑经历了各式各样的改建、润饰、大修，有的在意料之中，也有的处计划之外。这无损于阐释清晰、限定明确的整体结构，但毫无疑问会影响其中的某些组成部分。

最后还要强调，那些很多人认为堪称至高成就的东西，对鲍赞巴克来讲则是热切的，十载寒窗的苦学经历：漫漫路途中的跬步，而非旅程终点。

音乐博物馆的一部分，230座的宣讲厅，也可兼作音乐厅和电影院。
Part of the Music Museum, the 230-seat lecture theater also functions as a concert hall and cinema.

a number of striking deviations from the Conservatory and Cité approach. For ten years — 1984 to 1994 — he pursued his "major work" at Porte de Pantin, yet it would be a mistake to see his manner as no more than this. Nor his imagination, his sensuality, his enthusiasms and his explorations.

If any proof is needed, other projects offer a striking plurality/diversity/multiplicity of agendas and modes of expression.

Among them are the Café Beaubourg and the Musée Bourdelle, but also his EuroDisneyland hotel (1998), the Port de la Lune in Bordeaux and the housing units in Fukuoka, Japan (1989), the Crédit Lyonnais tower at Euralille (1992) — and others.

Monstrous, masterly and majestic, the Cité de la Musique, says its creator, owes as much to chance as to skill. At the outset the mix of genres made it unpredictable and hard to control, and the passing of time brought a multiplicity of modifications, retouchings and overhaulings, of approximations and fortuitous elements. This had no influence on an overall structure already defined with clarity and determination, but certainly affected some of its components.

In the final analysis, then, what for many is a crowning achievement remains, for Christian de Portzamparc, an intense, ten-year learning experience: a step along the way, rather than the end of a journey.

不可思议、古灵精怪：音乐城卵形的主音乐厅。
Surprising, strange: the oval main concert hall at the Cité de la Musique.

3

引燃
firing up

音乐中心，像一艘巨轮一般停泊在里约热内卢平坦的西部地区。

The Cidad da Musica, set like a great ship on the plain west of Rio de Janeiro.

远处科科瓦多地区起伏舞动的山脊线。

In the distance the rising, falling, dancing line of the Corcovado Range .

建筑之音

The music of architecture

沿索道和山路攀爬700余米，就到达了救世主耶稣圣像脚下，它也是科科瓦多山神之像。从这里俯瞰里约热内卢的海湾全景简直让人惊心动魄。不远处就是奇久卡国家公园的热带森林，那里散发着芳香，使得混迹于里约热内卢本地居民之中的你倍感欢愉，这些居民是科巴卡巴纳海滩和依巴内玛海滩的后裔子孙。

科科瓦多山摇摆舞动的身躯从这里蜿蜒伸展，山脊线连绵起伏，颇有超越自然力的气势，当你凝神注目，就会实实在在感受到这些群山在翩翩起舞——在来自五

Of course: the funicular and the 700-meter climb to the feet of the statue of Christ the Redeemer, the Christ of Corcovado. From up there the view over the bay of Rio de Janeiro is just breathtaking. And no distance away is the tropical forest of Tijuca, whose perfumes only intensify the euphoria you feel from being among the Cariocas, the people of Copacabana and Ipanema.

Out from there stretches the swaying, dancing line of the Corcovados. There's something weird about the way the ridges follow one another, and when you look hard you really do feel, straight away, that those mountains are dancing — that maybe they invented the samba here,

湖四海的人们开垦这片神灵庇佑的土地之前，或许正是这群山创造了桑巴舞。

山脚下，在大海与泻湖之间，奇久卡新区的规模与日剧增。

这里，在街区的心脏地带，音乐中心一定会建构起社区的结构和灵魂。这一力作包涵大小两座音乐厅，一所音乐学院，一所音乐学校和四座艺术电影院。

在里约热内卢，鲍赞巴克沉迷于某种巴洛克气质。这是对尼迈耶，对热带，对他的巴西籍妻子伊丽莎白，对阳光和广阔开放空间的颂词：这里有抒情风格与自发性的共鸣，有难以想像的充沛活力与迷人的轻巧，有简洁的平面和戏剧性的舞台调度，有典雅的视觉效果和挺厉的线条，还有曲线与反曲线，弯拱，穹顶与栏杆。不必再尽数描绘了，那么，交响乐厅效果怎样呢？巨硕的"花瓣"植根于地面，吐艳绽放，卷曲翻转，形成开口，提供广阔的视野与远景，鲍赞巴克对这些"花瓣"也有些面露难色；然而，巴西的混凝土工程师已经惯熟于尼迈耶粗犷的巴洛克风格，他们很快打消了鲍氏的顾虑：在这个新兴

long before people came from all over to settle this land blessed by the gods.

Down below, between the ocean and the lagoons, the Barratijuca neighborhood is growing faster and faster.

And here, in the very heart of the neighborhood, the Cidad da Musica is destined to provide structure and a soul. Quite a project, with a large and a small concert hall, a music school, a conservatory and four art cinemas.

In Rio, Portzamparc gave in to a certain Baroque temptation. A kind of tribute to Niemeyer, the tropics, his Brazilian wife Elizabeth, the sun, the wide open spaces: a perfect musical resonance combining lyricism and spontaneity, imaginative exuberance and lightness of touch, a simple plan with a dramatic mise en scène, visual elegance and linear clarity, curves and countercurves, arches, vaults and handrails. So much for the score, then, but what about the orchestra? Portzamparc was worried about his enormous folded "leaves", planted on the floor and opening out, scrolling in on themselves, providing openings, views and vistas; but the Brazilian concrete engineers, used to Niemeyer's Baroque boldness, were quick to put his mind at rest: in this new, totally receptive country nothing is impossible. A strange vessel, this

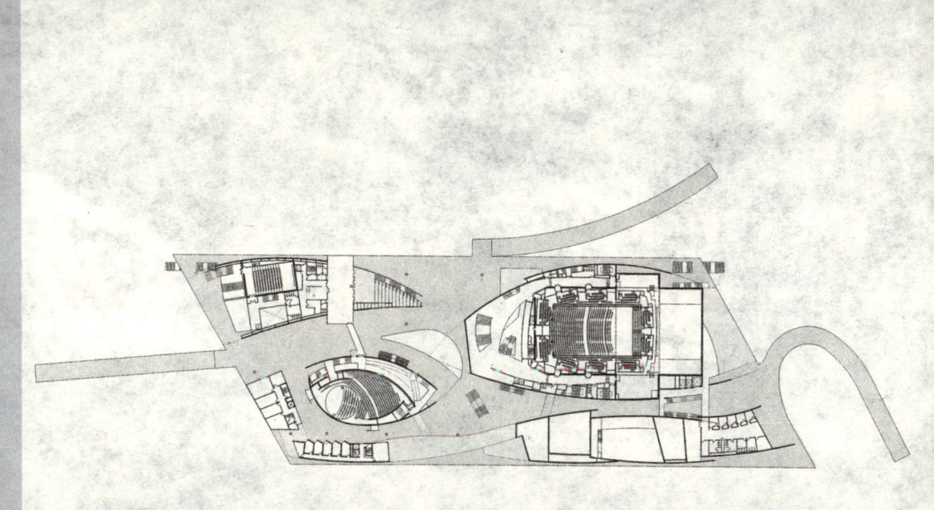

收进的艺术，片断化的设计：透明，联系，视野。

The art of setting-back, fragmenting: transparency, connections, vistas.

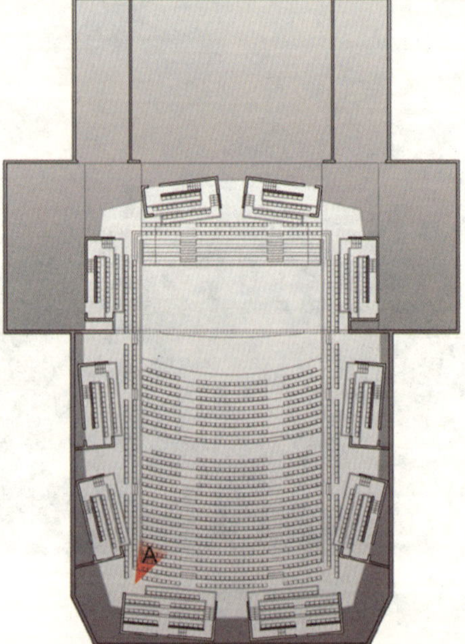

自成一景，主音乐厅被可沿滑轨运动的塔楼所环绕。

A spectacle in its own right, the main hall is ringed with towers that move on rails.

可移动的塔楼意味着音乐厅可分别举办交响乐和歌剧表演。

The movable towers mean the hall can function as a philharmonic or an operatic venue.

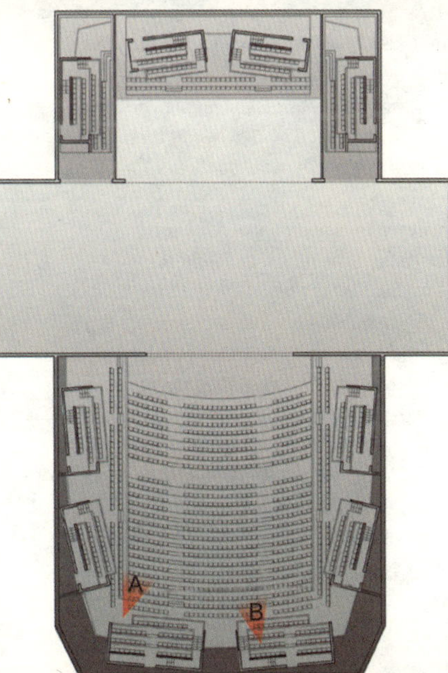
A
B

的，善于以他山之石攻玉的国度里，什么事都有可能发生。音乐中心形态奇异，所有的设计构思都旨在引导视线穿越建筑并延展至远方，将山川与汪洋尽收眼底，它就像一位来访者，脚下流淌着 10 余米的空隙，横跨过一座巨大的花园和一池庄重高贵的湖泊。

外部形体可以看作对尼迈耶的承袭，而主音乐厅无疑是对莎士比亚的礼拜——至少因袭了莎士比亚时期剧院的特征。800 个座位的布置采用交响乐模式，其余 1300 个座位则采纳歌剧布置模式，音乐厅周围环绕着容纳有包厢的独立体块，颇具鲍赞巴克的个人风范：观看与被观看，这

Cidade da Musica, where everything is designed to lead the eye through and out, taking in the mountains and the ocean, as the visitor, with ten meters of empty space under his feet, strides over a vast garden and a majestic lake.

The outside shape can be taken as a tribute to Niemeyer, but the main concert hall is unquestionably a homage to Shakespeare — or at least the theater of Shakespeare's time. With its eight hundred seats laid out on the philharmonic model, and thirteen hundred on the operatic model, the hall is ringed with towers housing boxes very much in the Portzamparc manner: see and be seen, the essence of the auditorium. What's more, the towers move on rails, disappearing when necessary to make room for a splendidly deep stage. Christian de Portzamparc is a music man, as everyone knows:

运动，节奏，开敞，光线。
Movement, rhythm, openings, light.

国立歌剧舞蹈学校：三种功能——舞蹈，学习，宿舍——统辖三个空间的设计。

The National Opera Dance School: three functions — dancing, studying, sleeping — governed the creation of three spaces.

是厅堂的精华之处。此外，这些体块可以沿导轨滑行，在必要时可以为深景舞台腾搁出空间来。鲍赞巴克是一位音乐人，众所周知：一位音乐家和爱好者。尽管只是业余音乐家——他曾涉猎钢琴和长笛——但他知识渊博，兴趣广泛，不仅敬仰马勒、德彪西、勋伯格、库塔、梅西安、利盖蒂，而且对巴赫、普赛尔、蒙特威尔地等人同样关注。更不要提爵士乐了——查特·贝克、迈尔斯·戴维斯、塞罗尼斯·蒙克，尤其是查理·帕克——当然还包括巴西的盖塔诺·维洛索、齐格·布瓦克等等。从埃里克·萨蒂音乐学校的“白色建筑”形制特征——这座原型式的坦比哀多位于

a musician and a music lover. An amateur musician — he's tried his hand at the piano and the flute — but a knowledgeable music lover, with a range of enthusiasms that set him afire as much for Bach, Purcell and Monteverdi as for Mahler, Debussy and Schönberg, or Kurtag, Messiaen and Ligeti. Not to mention jazz — Chet Baker, Miles Davis, Thelonious Monk and especially Charlie Parker — and the Brazilian sounds of Caetano Veloso, Chico Buarque and many others.

From what could be termed the "white architecture" of the Erik Satie Conservatory — that archetypal tempietto right in the heart of Paris' 7th arrondissement, where he goes back to the classics and tries out the fault-line as a basis for fragmentation — through to the lyrical flight

巴黎第七区的核心地带，鲍氏重拾古典，采用错落的线条创造破碎效果——到里约热内卢音乐厅焕发的抒情诗风韵，鲍赞巴克拓展了自己的视野，不知疲倦地探寻和希冀新的建筑和音乐领域。每一有迹可寻的变革中，他都会使用共鸣、谐音、甚至不谐和音等手法编排出独一无二的作品，这些作品在时空中逐步展开，一个乐章紧跟一个乐章地吹奏出你常常无法预计的倚音。

音乐，建筑：它们的关联显而易见。正像保罗·瓦莱里在《欧帕里诺斯》(《Eupalinos》)中坚称："告诉我，对建筑灵敏善感的人啊，难道你没有注意到，当你穿行于城市之间，鳞次栉比的建筑中，有些哑口无言，有些则侃侃而谈，而极有限的几座在引吭高歌？"(保罗·瓦莱里，《欧帕里诺斯或建筑》，格利玛出版社，巴黎，1921年。Paul Valéry, Eupalinos ou l'Architecte, Gallimard, Paris, 1921年)

楠泰尔：螺旋形主楼梯及其平台是偶遇与交流的理想场所。

Nanterre: the main spiral staircase and its landings are points for encounter and interchange.

of Rio, Portzamparc has been broadening his scope, ceaselessly prospecting new architectural and musical territories. And with each seeming change of register, he uses consonance, assonance and even dissonance to compose a unique work, one unfolding through time and space and embracing, movement by movement, appoggiaturas you don't always anticipate.

Music, architecture: the connection is obvious. As is the recourse to Paul Valéry, in his Eupalinos: "Tell me, you who are so sensitive to architecture: haven't you noticed, as you walk around the city, that of the buildings that populate it some are mute, others speak and still others, the smallest in number, actually sing?" (Paul Valéry, *Eupalinos ou l'Architecte,* Gallimard, Paris, 1921)

If it's true that the body experiences music as time, we can only agree with Bernard Sève when he says, "All the arts are arts of time and all experience is the experience, indirect at

如果身体将音乐作为时间来体验，我们就只能认同伯纳德·塞维的说法，"一切艺术都是时间的艺术，一切体验都至少间接地具有时间性。"（伯纳德·塞维，《音乐的堕落》，塞伊出版社，巴黎，2003年。Bernard Sève，L'altération musicale，Le Seuil，Paris，2003年）塞维谈及的是即时的、隐晦难解、不可度量的时间，"游离于外部目标和阈限之外"。在他音乐般的创作中，鲍赞巴克将音乐化的时间塑造为空间的基本元素，作为其空间化过程。

因此音乐成为浓缩于空间之中的时间，同理舞蹈可比拟为空间中提炼出的运动：无论从个人还是从职业角度思考，舞蹈都是鲍氏青睐的另一重要主题。

鲍氏身上带有几分默斯·康宁安的

least, of temporality." (Bernard Sève, *L'altération musicale*, Le Seuil, Paris, 2003) Sève is talking about an immediate, impalpable, unquantifiable time, "freed if external goals and constraints". In all his musical creations, Portzamparc turns this dimension of musical time into a fundamental element of space and the process of spatialisation.

Thus music becomes time compressed within space, in the same way that dance can be taken as movement drawn out in space: for dance, too, is an essential aspect of Christian de Portzamparc, personally and professionally.

There's something of Merce Cunningham about him. In the silhouette, the "forever young" walk, the distinctive way of being absolutely classical and unfailingly contemporary, the determination to endlessly renew the vocabulary in its use of gesture, movement and density, in its dazzlingness and appropriation of space. Each

德彪西对斯特拉文斯基的评价："亲爱的挚友，你对未知领域的拓垦使我欣慰不已。"在楠泰尔，鲍赞巴克如法炮制。

Debussy to Stravinsky: "Dear friend, it gives me intense pleasure to see how you push back the frontiers." At Nanterre, Portzamparc does the same.

气质。这种相似性仿佛描模一般惟妙惟肖——一样"青春永驻"的步伐，一样地道的古典遗风与经久不衰的现代韵味的完美结合，一样坚韧不拔地运用举止姿态，运动和密度（等手段）更新艺术语汇，而创作出的空间一样气宇轩昂、舒适宜人。这一切旗帜鲜明地重述了德彪西对斯特拉文斯基的评价："亲爱的挚友，你对未知领域的拓垦使我欣慰不已。"1983 年，鲍赞巴克在楠泰尔巴黎歌剧舞蹈学校竞赛中夺魁，四年以后，他建造了一处"破碎化"的场所，其中的三个主要空间容纳匹配了三种功能：舞蹈（空间），学习（空间），宿舍（空间）。这一切活灵活现地预示了即将被他采用的设计模式：序列的创造，声学独立体，光线的感应，冲突与对立的激发，正像这座建筑的螺旋形楼梯及其平台所展示的那样。在规避古典主义的同时，他勾勒出后来在潘廷门音乐学校中运用的种种基元。

在楠泰尔，"田鼠"们——学生们的别号——三人同室而眠。他们年龄各异：这有益于在芭蕾舞团中建立团队精神，也方便年轻的舞者向学长请教。一位女生——约摸 13 岁左右——称赞恭维了建筑师的作品，但却有一件遗憾之事，她告诉建筑师：在加尼耶歌剧院幽长昏暗的走廊中，她曾因恐惧而微微颤抖，这感觉美妙奇异，但楠泰尔却失掉了这种氛围。

在楠泰尔对光的主题探究一番之后，鲍赞巴克进而研习浓密的阴影，他重读瓦莱里，但却拥有更为犀利的眼光："让舞者足登狭紧的舞靴；这样，他们将演艺出崭新的舞步。"

1983 年，他在巴士底歌剧院竞赛中兵败垂城，尽管他的方案出类拔萃，臻于完善，因地制宜，但却由于一条华而不实的

1983 年巴士底歌剧院方案石沉大海。不公正？
The Bastille Opera project failed to win in 1983. Unjustly?

方正的基地上嵌入灵动的斜线：鲍氏的方案"撰写"了巴士底广场。
An oblique dynamic for a square space: the Portzamparc project "wrote" the Place de la Bastille.

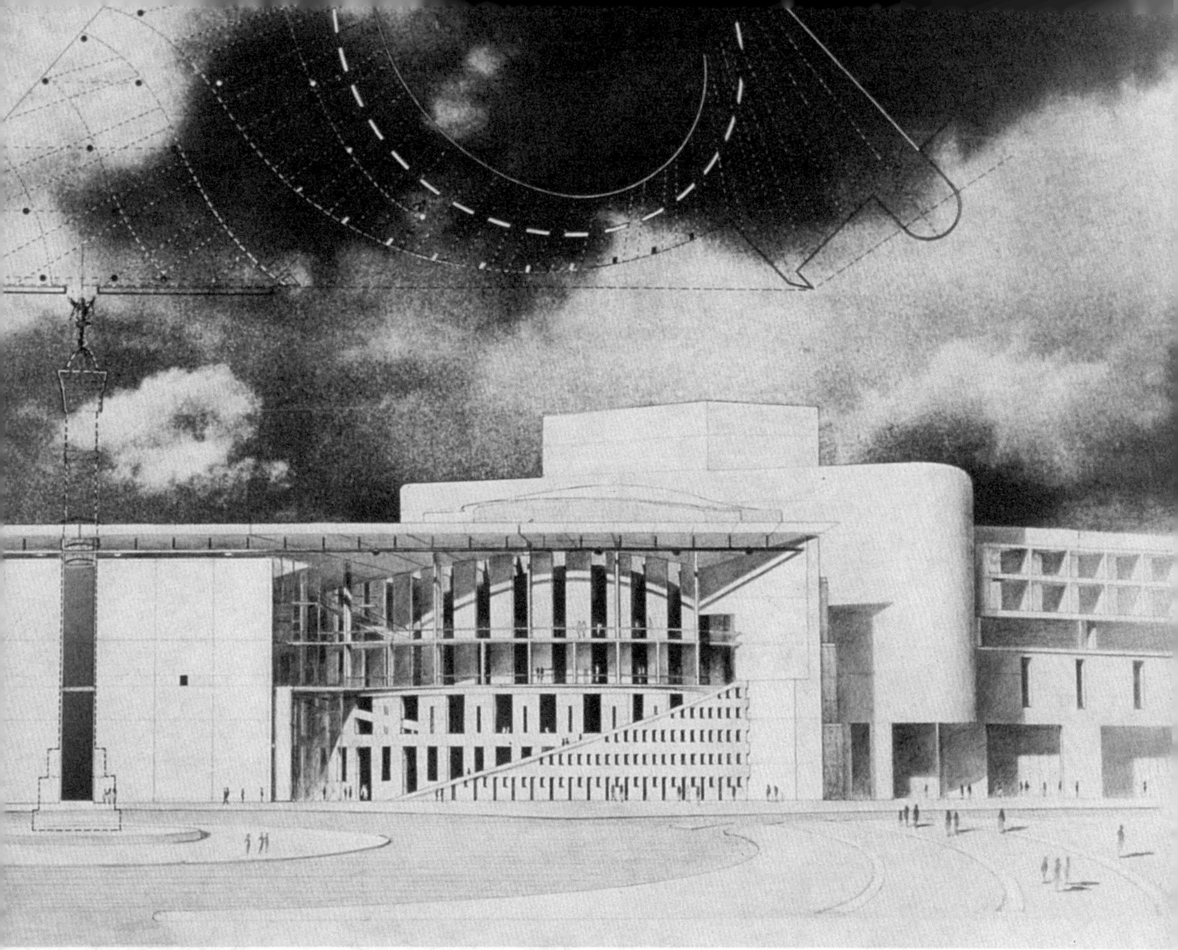

of them is vibrant with Debussy's remark to Stravinsky: "Dear friend, it gives me intense pleasure to see how you push back the frontiers." In 1983 Christian de Portzamparc took out the competition for the Paris Opera Dance School at Nanterre. Four years later he delivered a "fragmented" site whose three main spaces fit with three functions: dancing, studying and sleeping. This was a thoroughgoing introduction to his coming manner: generation of sequences, acoustic separation, induction of light and encouragement of encounters, as on the great spiral staircase and its landings. At the same time as he was abandoning classicism, he was outlining what would later be the Conservatory at Porte de Pantin.

At Nanterre the "rats", as the students are called, sleep three to a room. The three are different ages: this helps create an esprit de corps within the corps de ballet and means the young dancers learn from their seniors. One of the girls — she might be thirteen — compliments the architect on his work, but there's one thing she misses, she tells him: the deliciously magic frisson of fear she felt in the long, gloomy corridors of the Opéra Garnier.

After discovering light at Nanterre, Portzamparc moved on to dense shadow, reading Valéry again, but with a sharper eye: "Make the dancers wear shoes too small for them; that way they'll invent new steps."

This was in 1983, the year when he lost the

建造法规而最终落选。有一派势力曾保举理查德·迈耶，但最终花落卡洛斯·奥特之手。巴士底广场至今仍未修葺；多年之后，当密特朗接受电视采访时被问及“您有什么遗憾之事”时，他回答道“是的，鲍赞巴克的巴士底歌剧院”，多么诚恳的感言！

在巴士底方案中，鲍氏对城市进行反

Opéra Bastille competition over a specious building regulations objection, even though his project was by far the best, the most accomplished and the most appropriate. One school of thought wanted Richard Meier, but in the end it was Carlos Ott. The Place de la Bastille has still not recovered; and many years later, on television, François Mitterrand, when asked, "Do you have any regrets?", replied "Yes, the Portzamparc Opéra Bastille". Duly noted.

滑动幕墙的象征主义造就了强健的立面，浑然一体的破碎化（效果），和其他种种内涵。

The symbolism of the sliding curtain gives rise to a powerful facade, unifying fragmentation and all sorts of interpretations.

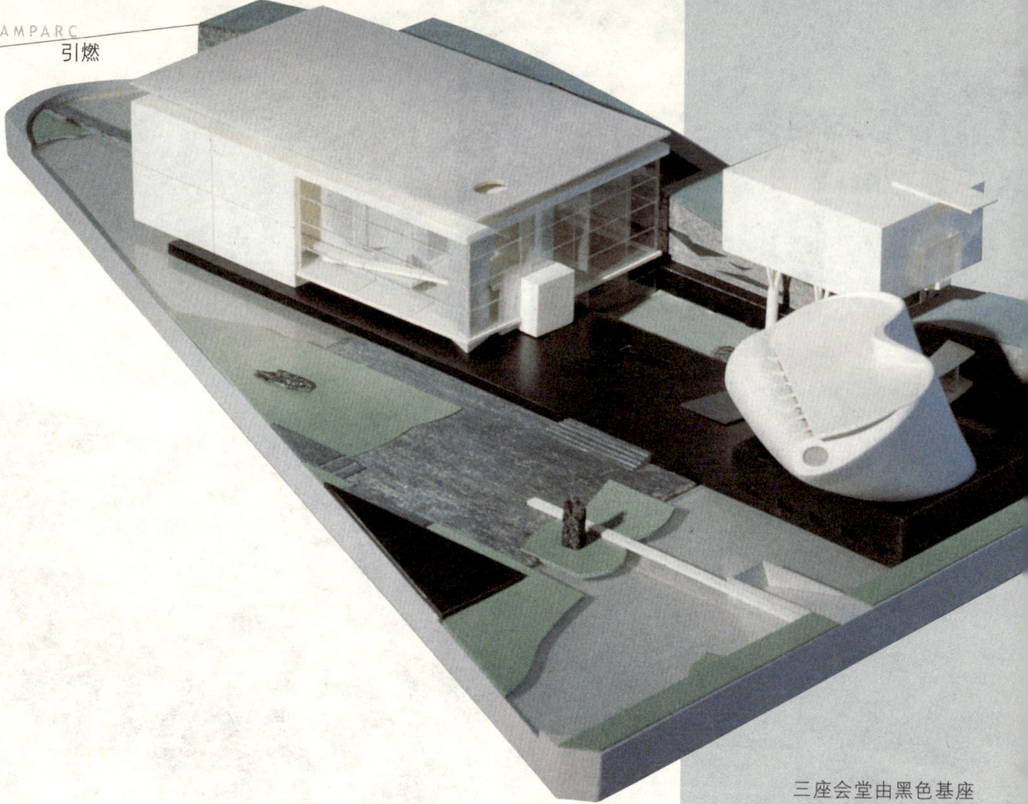

三座会堂由黑色基座统帅，创造了与奈良完美结合的建筑景观。
Three halls unified by the same black base offer an architectural landscape in perfect harmony with Nara.

思，古典的空间能否找寻到当代的形式表达？现代建筑——生来便与古典城市分道扬镳——能否重新融入城市脉络？混沌晦涩的构图为巴士底广场赋予了深厚的意义与象征："七月纪念碑"，圣-安东尼城郊街，对巴士底狱自身的记忆，圣马丁运河。广场巍然端坐，历史沧桑，气韵非凡，然而却缺少能表达凭吊之情的碑刻铭文。

鲍氏在方正的基地上嵌入灵动的斜线，他的方案"撰写"了巴士底广场。尤可感叹的是它还创造了新颖的象征主义：可滑动的幕墙孕育了复杂多变的内涵；建筑还拥有强健的立面；以及浑然一体的破碎化效果。

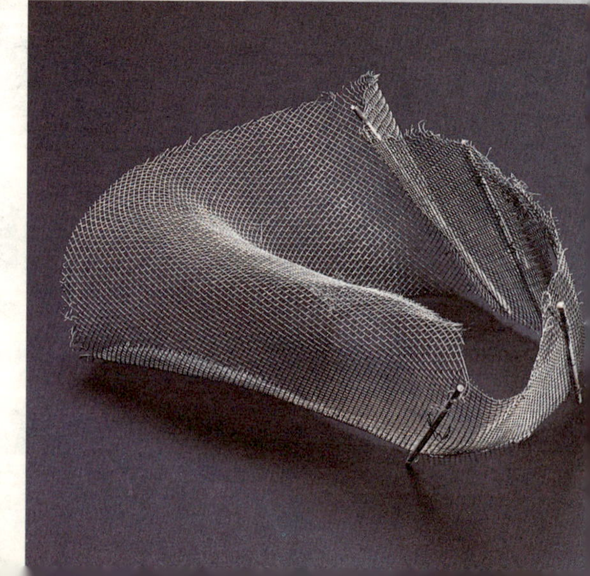

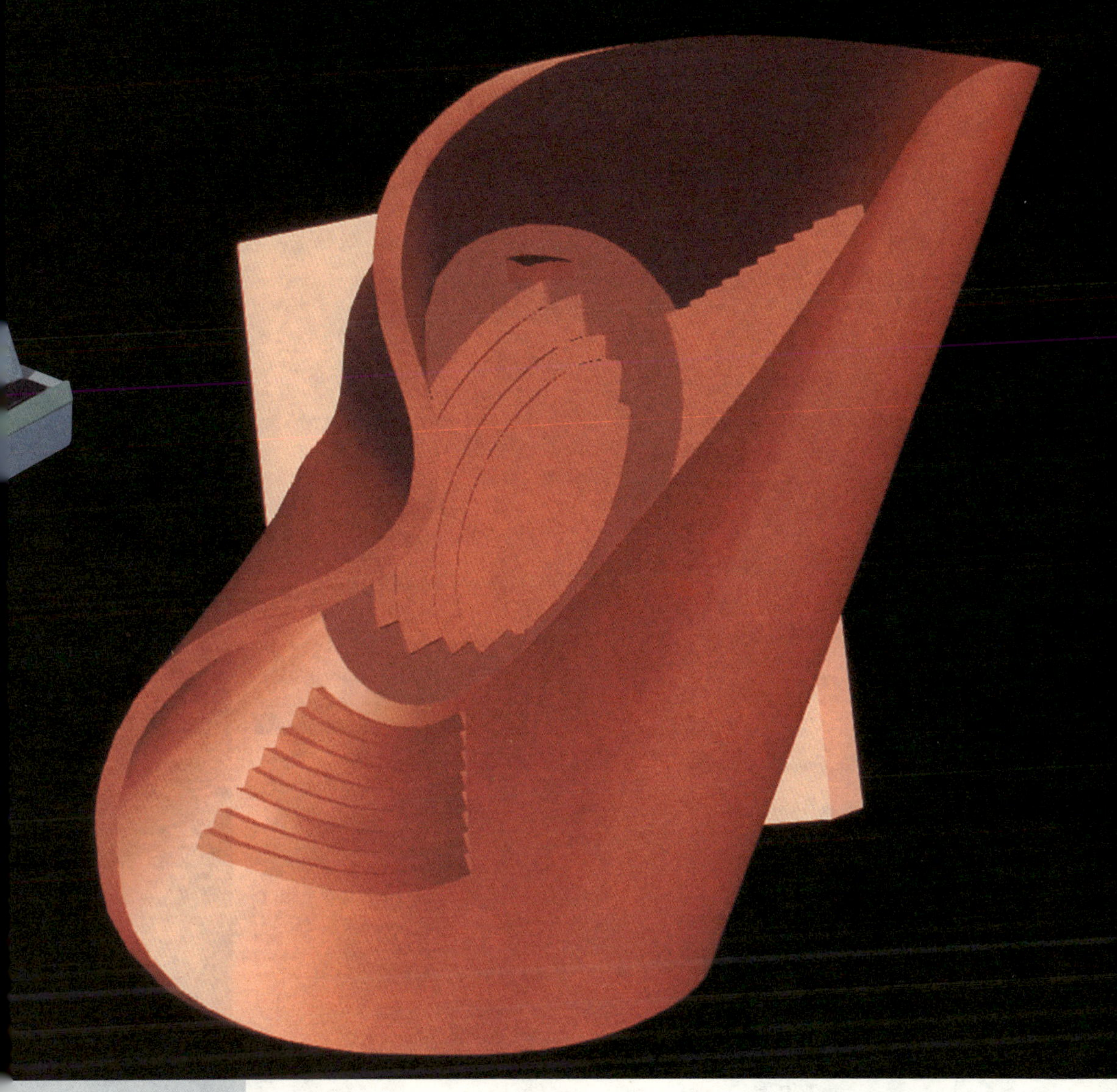

奈良：鲍赞巴克探索莫比乌斯带的可能性与奥妙之处。

Nara: Portzamparc explores the possibilities and subtleties of the Moebius strip.

In his Bastille project Portzamparc is reflecting on the city. Can classical spaces find a contemporary form of expression? Can modern architecture, born of opposition to the classical city, return to that city? A kind of chaotic working drawing, the Place de la Bastille is terribly weighted with meaning and symbol: the July column, the Faubourg Saint-Antoine, the memory of the Bastille itself, the Saint-Martin canal. The Place was there historically and symbolically, but it lacked inscription.

Inserting a dynamics of obliqueness into a square site, the Portzamparc project "wrote" the Place de la Bastille. All the more so in that a fresh symbolism was created: that of the sliding stage curtain and its varying interpretations; of the forceful facade; and of a unifying fragmentation. The following year Christian de Portzamparc found (partial) consolation for this failure by winning the competition for the Cité de la

第二年，鲍赞巴克赢得了音乐城竞赛，这使他获得了些许抚慰。他对音乐和舞蹈的志趣也并未就此截至。

1992 年，他参与了日本奈良国际会堂的设计竞赛，项目要求在与老城接壤的新区当中设计一组由三座礼堂组成的建筑群。奈良的公共庭园是古代庙宇的桃花源，其中一些为中（国）式，其历史可追溯到公元 9 世纪。鲍赞巴克直接从场地吸纳灵感，规划了三座颇具雕塑感的独立"白色建筑"，它们被公共中厅连接起来，坐落在黑色岩石基座上。

方庭，铺装地面，池塘与植被，交织为自成一体的建筑景观。这一综合体不仅承担娱乐和会议功能，也是供人冥想的场所，对历史久远的日本颇具慧眼的颂词，同时于现代气韵又毫发无损。

这一"历险"见证了建筑师对两类未知领域的探索：他的事务所运用出神入化的 3D 技术；还有莫比乌斯带，他长久迷恋于这一几何形态，因为它可以创造出没有间断线条的空间，人们能够同时身处室内与室外。

在奈良，他将著名的莫比乌斯带撕裂开来，插入采光口，在曲率最大处用玻璃顶封闭曲面，后来的事实说明，莫比乌斯带——使人回忆起葛饰北斋的"巨浪"，并联想到冲浪板的轨迹——以独特的方式成为卢森堡基尔勃格室内乐堂的某种预兆。

1994 年，鲍氏的事务所在另一竞赛中遭遇滑铁卢——哥本哈根文化音乐中心。

Musique. Nor was his foray into music and dance to end there.

In 1992 he entered the competition for the International Convention Center in Nara, Japan, a brief calling for a group of three auditoriums in the new district bordering on the old city. Nara public gardens are home to ancient temples, some of which are Chinese and date from the 9th century. Taking his inspiration directly from the site, Portzamparc proposed a group of three separate, sculptural, white buildings linked by a lobby and set on a base of black stone. Courtyards, paved areas, ponds and plantations combined to form a garden that is an architectural landscape in its own right. While an entertainment and congress venue, the complex was also a place for meditation, a vibrant tribute by the architect to age-old Japan in a style that, yet again, was uncompromisingly contemporary.

This venture saw the architect trying out two kinds of fresh terrain. The 3D technology his agency now used with such virtuosity; and the Moebius strip, whose geometry had long fascinated him in that it enables the creation of a space containing no broken lines and in which one is simultaneously indoors and outdoors.

In Nara he tweaked the famous ring a fraction by introducing a light-breach, a small glass lid that closes the strip on its tightest curve. As later became clear, the Moebius strip — an evocation both of Hokusai's "great wave" and of the trajectory of a surfboard — prefigures in its own way the Kirchberg chamber music concert hall in Luxembourg.

In 1994 the Portzamparc workshop lost another competition, for the Copenhagen Cultural and Concert Center. This provided, nonetheless, an opportunity to explore the overlaying of two levels devoted to different activities: a matter

在哥本哈根，建筑师实践双层叠加形式——这一手法后来在汉城，蒙特利尔，雷恩和里约热内卢反复出现。

In Copenhagen the architect explores two superimposed levels — an approach that reappears in Seoul, Montreal, Rennes and Rio de Janeiro.

然而，这提供了一个研究双层叠加形式的机遇，不同的自然层服务于不同的活动与行为：它关涉"浮力"问题，是其他实践项目的先驱——在巴黎的马约门，在汉城，在蒙特利尔，在雷恩和里约热内卢。鲍赞巴克不断重复和更新整体性设计手法、运用和实验各类风格体裁、主题变奏与破碎化效果，引领我们——音乐与舞蹈也不离左右——进入美国舞蹈编排艺术家崔沙·布朗的结构性即席创作理念："我着力实践，"她解释道，"因为它在空间中赋予你定位，同时也包含体量。"她的即席创作来源于对序列的处理；舞者们占据其核心位置，各自身处在假想的立方体空间中，显示出假想的体量。空间节奏，即时性的剔除，对未知领域的探索：正是这些将舞蹈，音乐与建筑熔铸归一。

前文曾叙述过与里约热内卢项目息息相关的抒情式巴洛克风格。而眼下在卢森堡基尔勃格高原上建造的音乐厅则与传统

of "buoyancy" that was the forerunner of other adventures in Paris (at Porte Maillot), Seoul, Montreal, Rennes and Rio de Janeiro. Endlessly repeated and renewed, the overall approach, stylistic devices, variations on a theme and fragmentation lead us — music and dance never being far away — to the American choreographer Trisha Brown and her concept of structured improvisation: "I'm trying it out," she explained, "because it situates you in space while including volume." Her improvisations are based on work with line; at their core are the dancers, each in a space of his own conceived as a cube and so revealing an imaginary volume.
Spatial scansion, abolition of temporality, pushing back of frontiers: these are what unify dance, music and architecture.
Mention was made above, in connection with the Rio de Janeiro project, of a kind of lyrical Baroque. But what is currently being built on the Kirchberg plateau in Luxembourg is closer to classical Baroque. Two concert halls in the same complex: one on philharmonic lines, with seating for 1500 and the other, for chamber music, offering space for 300.

在卢森堡再次出现鲍赞巴克式的主题：柱林和锥体。

Recurrent Portzamparc motifs in Luxembourg: the forest of pillars and the cone.

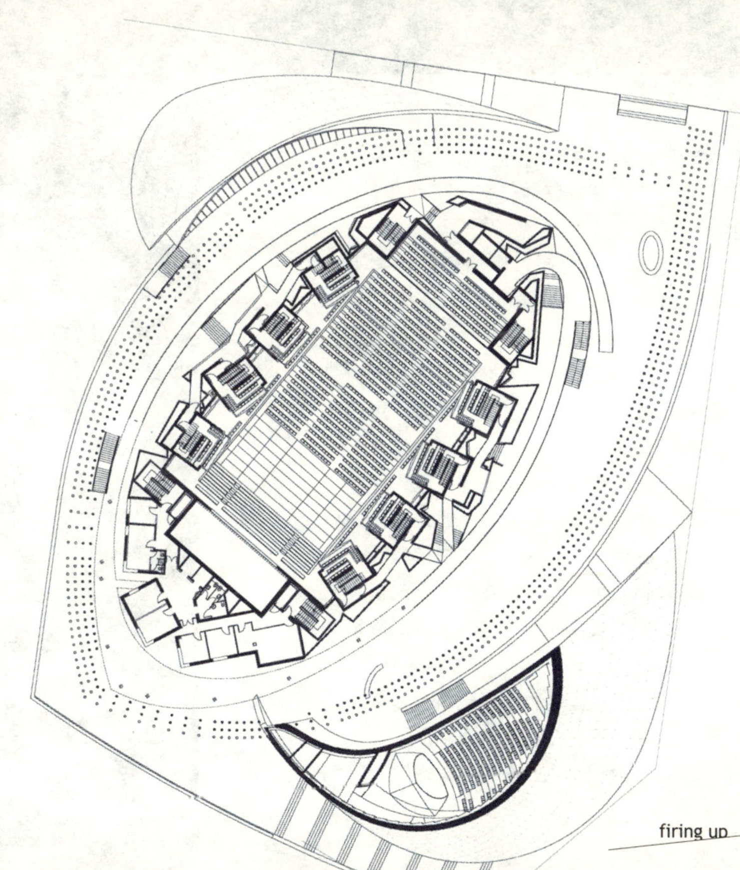

1500 席的交响乐厅和
300 席的室内乐乐厅。

Coexistence:
a philharmonic hall seating 1500, a chamber music room for 300.

巴洛克风格更加神似。同一综合体中设置两座音乐厅：1500 席的交响乐厅和 300 席的室内乐乐厅。

与拉维莱特大音乐厅的椭圆形态不同，在这里，17 米高的平行六面体具有不规则的外墙和 8 尊布置了多层包厢的小型塔楼。塔楼的轮廓线显得怪异嶙峋——像断崖一般，它们塑造出极尽动感、声色了然的紊乱效果。

小音乐厅在同一标高上与交响乐厅贴近布置，这是继奈良会堂后，运用莫比乌斯带的又一版本。在这里，曲面如树叶一般铺展开去，蜷曲弯转，然后又挺阔直立起来。人们再次感受到情绪与感官的激荡沉浮，在这座逾越规矩的完整的建筑中，类似的效果比比皆是。这座建筑并非源自场所的秉赋，却应归功于音乐的恩赐（灵感），尽管场地和周边环境促使鲍赞巴克

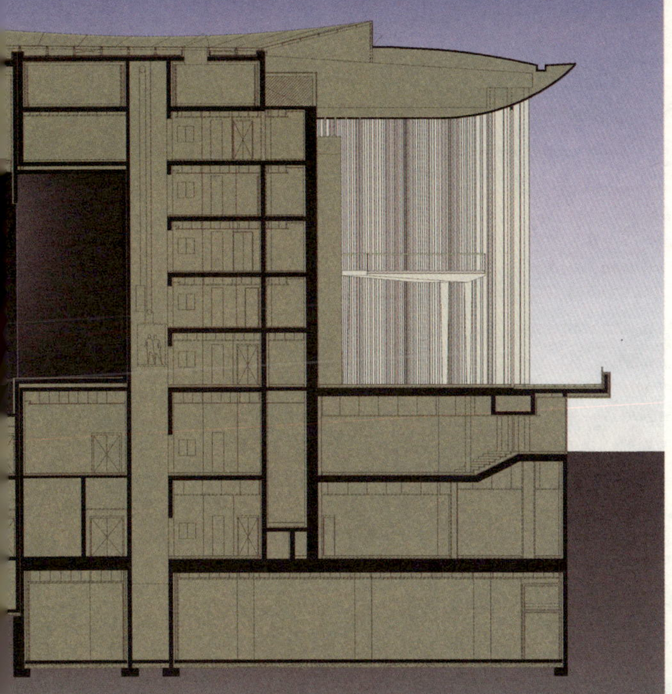

At La Villette the large hall is oval. Here, 17 meters high, it's a kind of parallelepiped with irregular walls and eight little towers housing tiered boxes. The actual contours of the towers seem oddly ill-defined and, like cliff-faces, they generate an extraordinarily moving, sensual disorderliness.

Set close by at the same level, the small hall is a version of the Moebius strip as planned for Nara. Here the strip becomes a leaf that opens out, twists, lies flat and straightens up again. Once again, one feels emotion and sensuality everywhere, which is hardly surprising when the building, an exception to the rule, is considered as a whole. It is a building born not out of the nature of the site but solely out of music, even if the site and its built environment have clearly led Portzamparc to emphasize a truly exceptional shape and treat the brief as calling for the creation of a jewel in its own setting.

主音乐厅中断崖般的塔楼，塑造出极尽动感、声色了然的氛围。

The cliff-like towers in the main hall generate a movingly sensual atmosphere.

使用别出心裁的形式，将创造归属于场地自身的建筑瑰宝作为第一要务。

最终成果是一座略失规整的椭圆建筑，林林总总的柱林环绕其侧，如同光线穿透镜片一般，产生了某种过滤效果。究其根本，这椭圆形体出自巴洛克风格；然而它却时时表达着对"黑暗中世纪"的拒斥和对光明的无限向往。它翻腾雀跃，波动流连，艳丽多姿的空间组合引导人们踩着优雅的步伐摇摇地漫步其间，甚至让我们觉得这空间出自雕塑家之手。

巴洛克是表现的艺术：在这里，我们能做的就是沉浸于它的节奏与起伏之间，椭圆与幻象之中，英雄主义与声色犬马之内，奢逸艳丽与奇巧的含混性之侧。

The outcome is a slightly irregular oval ringed by a forest of piers that make up a kind of filter — something to be "passed through", like a looking glass. An oval which, naturally, comes out of the Baroque; and which cannot fail to remind us that the Baroque and the opera were born together out of a rejection of "the dark ages" and an aspiring to enlightenment. An oval whose billowings, undulations and voluptuous spatial combinations so clearly evoke a graceful, swaying walk even as they give the impression that the space has been hollowed out by the hand of a sculptor.

The Baroque is an art of presence: all we can do here is surrender to its interplay of rhythm and undulation, ellipse and illusion, heroism and sensuality, voluptuous ecstasy and deft ambiguity.

卵形，椭圆，螺旋，幻象，曲线，反曲线：地道的巴洛克语汇。

Ovals, ellipses, whorls, illusions, curves, countercurves: a totally Baroque vocabulary.

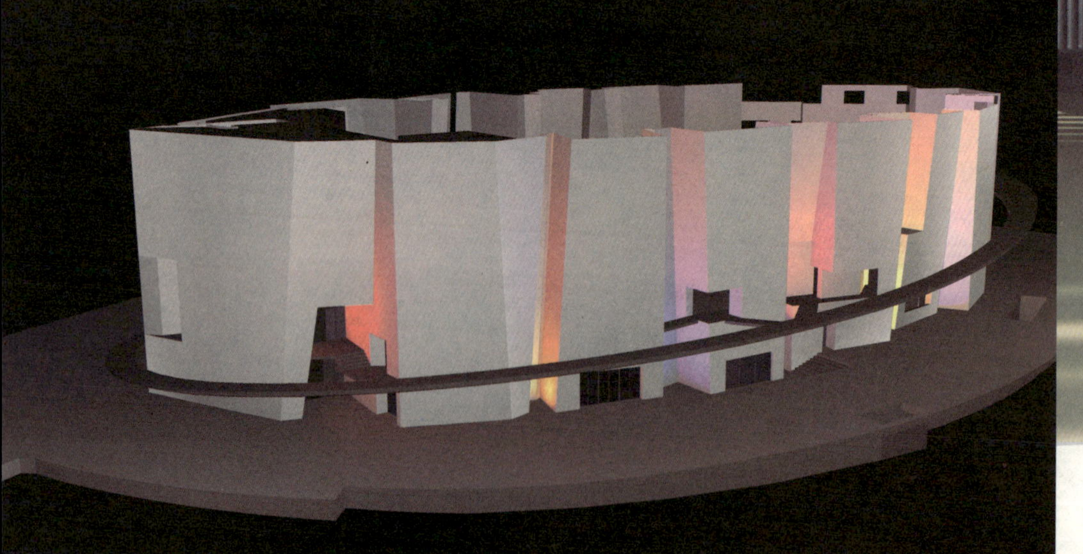

与拉维莱特和"万代"(Bandai)的项目相比，在这里，颜色起到至关重要的作用。
Even more so than at La Villette and in the project for Bandai, color plays an essential part here.

室内乐乐厅被比作一片铺展开去，蜷曲弯转，然后又挺阔直立起来的树叶。

The chamber music room is seen as a leaf that opens out, twists, lies flat and straightens up again.

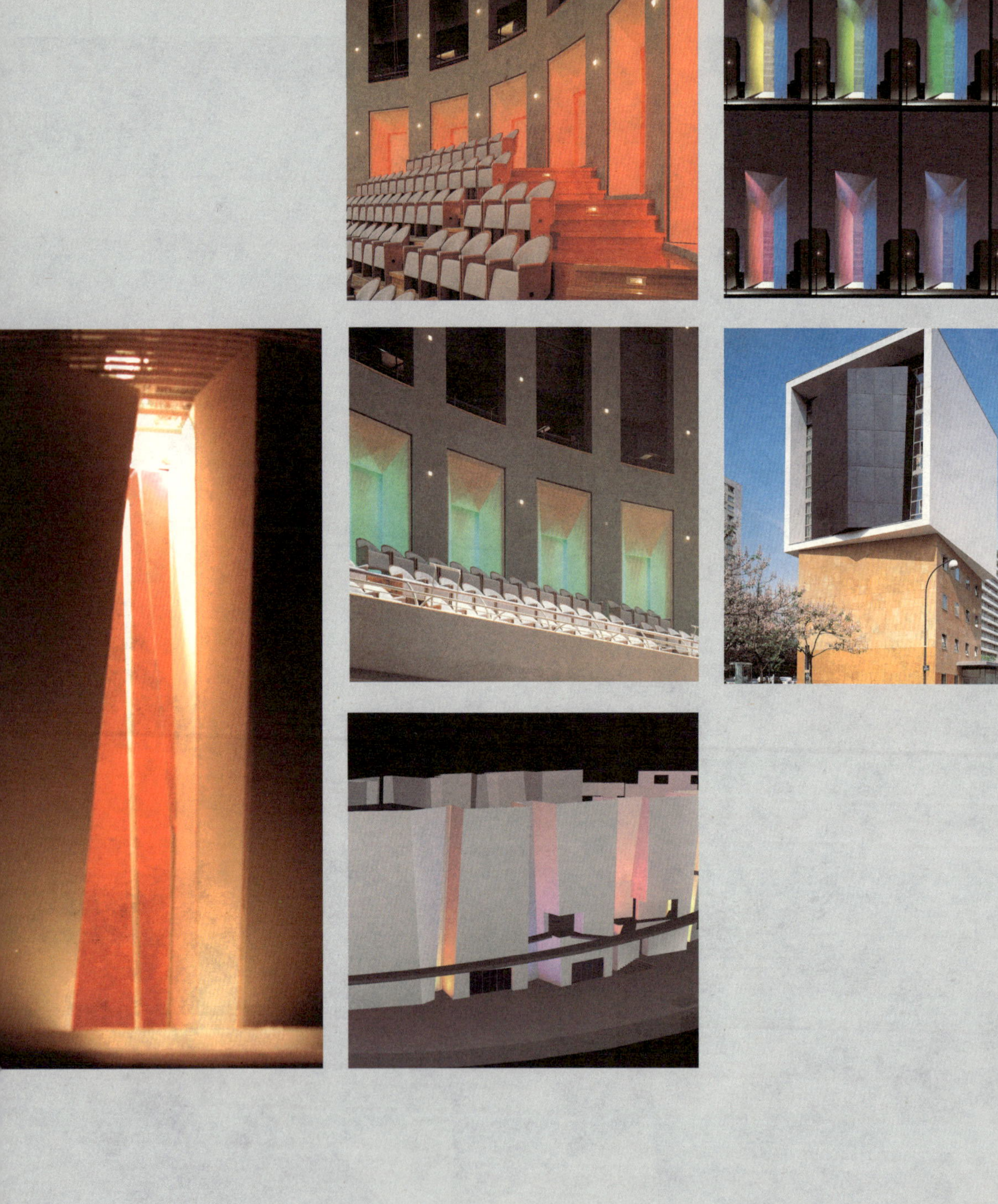

TUSCHINSKI

4

伸张

reaching out

自 1979 年的欧风路项目始，鲍赞巴克一直致力于诠释他的规划理念与建筑信条：景观，破碎，交叉，光线。

Since the Rue des Hautes Formes in 1979, Portzamparc has been displaying his planning and architectural credo: landscape, fragmentation, crossings, light.

开放式街区

The open block

鲍赞巴克早期是勒·柯布西耶的追随者。

当1962年他开始在巴黎美术学院的学业时，他是——至少自封为——忠实的"柯布主义者"和彻底的"现代主义者"。晚些时候，他参与了几个学术团体，参与的活动包括泛读文献和"学术报告"，而宗旨是为所有人创造更为惬意的场所。于是乌托邦理想隐隐呈现，这理想一如既往地凝重繁复……

阅读吉尔·拉布吉关于希波达莫斯——他被希腊后裔们尊为真正意义城市规划的创立者——的评论，鲍氏认识到"一人同时扮演两种角色——建筑师和乌托邦理想主义者——并非出于机缘巧合。"拉布吉强调，这触发了一场艰苦卓绝的争斗——有机与秩序的对峙——一场鱼贯整个人类发展历程的辩论。他还补充道："乌

In the beginning was Le Corbusier.

When Christian de Portzamparc began art school in Paris in 1962 he was — or at least he thought he was — very "Corbusian" and utterly "modernist". Later he would join study groups that involved extensive reading, "reportings" and generally making the world a better place for all. So much so that Utopia began to loom, burdensome as ever...

Reading Gilles Lapouge on Hippodamos, credited by the later Greeks with the invention of formal city planning, he learnt that "There was nothing fortuitous about the two functions — that of architect and that of Utopian — emerging simultaneously in the same man." Emphasizing that this was the beginning of the violent quarrel — organic versus organized — that would mark humanity's subsequent peregrinations, Lapouge adds: "The Utopian makes human life and the life of human societies the subject of an equation. His

之前 / 之后
before/after

在国家大道，他创造性地对现状建筑进行城市重组与复原。
With the Rue Nationale, he combines urban recomposition, rehabilitation of existing buildings and architectural creativity.

analysis of the perfect, atemporal, anonymous world of numbers is aimed at making man into an atemporal, anonymous, perfect object. He elucidates the system of the cosmos, not to ensure man's sovereignty over it, but to turn man himself into a system. He subordinates all anomalies to the norms of the plan. He reorganizes unpredictability as monotony, disorder as regularity, chance as logic. He casts desire in the mould of need. He will see liberty extinguished that equality might triumph. He is a structure fanatic whose dream it is to inject structure into the life of individuals, of societies, of peoples..." (Gilles Lapouge, *Utopie et Civilisations*, Paris, Weber, 1973)

Portzamparc shuddered, pondered, hesitated, sensing a new Utopia, a new barbarism in the wind: that of technico-economic tyranny, of territories defined in terms of GDP and GNP. He had already glimpsed the issue of which Olivier Rolin — to be a fellow Maoist in 1968 — would write forty years later: "Barbarism detests beauty; it is one of the most abhorrent outcomes of collective ambition." (Olivier Rolin, *Tigre en papier*, Paris, Le Seuil, 2002) In 1966 he took a break, cut free of architecture and left for New York, where he discovered fresh literary, musical and life-style paradises. But he went on thinking, too, and finally, literally, "reacted" his way back to architecture. When it came to the crunch, he was infinitely more Latin than he thought.

Back to the sources, then. Starting from scratch. Bramante, maybe? Bramante in Vigevano, that is: in Lombardy, where he drew an curiously oblong residential square, whose arcades underpinned houses aligned every which way and decorated with painted motifs.

It resembled, in fact, any number of those Italian squares so conducive to reverie, intimacy, encounters, relaxedness. Returning to Lapouge, he found: "The anti-Utopian is a man of passion. His specialty is not the real, but the desirable. With no time for logic, he makes it a point of honor to humiliate good sense, reality and reason. The man of history is a dialectician, less drawn to the designing of structures than to the interactions between events as the process takes its course. He is a man of metamorphoses.

托邦主义者将人生与社会生活当作方程式来处理。他们对完美、非时间范畴、不具名的数字世界孜孜钻研，试图将人类整合为非时间范畴、不具名、完美的客体。他们阐释宇宙的系统，并非意图使人类孑然独立，而是要将其熔融在系统之中，他们置一切不规则于拟就的规范之内，将无可预料之事归于千篇一律，把凌乱合入规则之中，化随机于逻辑之内。他们将欲求浇铸在‘需要’的模具中，见证了自由的泯灭与均等性的凯旋。他们是结构与组织的信徒，梦想将私人生活、社会与人类活动赋予结构性……”（Gilles Lapouge，Utopie et Civilisations，Paris，Weber，1973 年）

鲍赞巴克战栗着、沉思着、踌躇着，

之前 / 之后

before / after

细细体味着全新的乌托邦理想，那是一种凛然而至的野蛮与暴虐：技术和经济的独裁，由GDP和GNP指数统帅的领地。他已瞥见奥利维尔·罗林（Olivier Rolin）40年后写下的论点："野蛮憎恨美；它是勃勃野心聚积郁结而成的最格格不入的毒果。"（奥利维尔·罗林，《纸老虎》，塞伊出版社，2002年；Olivier Rolin，Tigre en papier，Paris，Le Seuil，2002年）1966年，鲍赞巴克稍事休整，暂停建筑工作，动身赴纽约，在那里，他发现了生机勃发的新文学，进入了音乐与美妙生活方式的伊甸园。但他沉思不殆，最终真切地作出"反馈"并回归建筑。在关键时刻，鲍氏拉丁化的程度远比他自认的程度为甚。

让我们回归源头，从起点开始探究。也许一切都源自伯拉孟特？维杰瓦诺（Vigevano，地名——译者注）的伯拉孟特：他曾在伦巴第设计了一座奇妙的矩形市民广场，拱廊庇护着成排的住宅，并装饰有各类主题的壁画。

与意大利的其他许多广场相似，它传达了对白日梦式的幻想、亲密与舒适感、不期而遇和休闲松弛气氛的仰慕欣赏之情。重提拉布吉的文字，鲍氏发觉："反乌托邦理想的人均是满腹激情之士。他们的着眼点并非现实，而是欲求。忽视逻辑的存在，甚至以蔑视现实、理性、意义为荣。抱有历史情结的人是辩证主义者，不

The Utopian, by contrast, has no dialectical sense whatever. What use could he make of it? He hates movement, because it modifies line. He is a logician." (Gilles Lapouge, op. cit.)

His return was towards Europe, France and Italy, with Bramante and Vigevano at the top of his list. "What Italy abhors," he read in a book by Dominique Fernandez (Dominique Fernandez, *Civilisation latine*, Paris, Olivier Orban, 1986), "is the big drafty French-style square, a place to pass through rather than one of refuge, recovery, close contact. No Roman stops on the Piazza del Popolo, because you can't pursue an affectionate daydream there. All you can do is traverse. It's like the Place de la Concorde in Paris, the archetypal drafty square and totally alien to the Italian spirit. Why this desire for an enclosed place? Because the Italian square is a place for living. In France you live at home, indoors. When you go out, it's to go to another indoors. In a Latin country, you don't live at home, you live outdoors; but to be a place for living this outdoors has to look like an indoors. The advantage of the square over the house is that you're together. The Latin can't stand being stuck at home, it makes him feel cut off, excluded from the company of men and women. That's the essence of it. The square responds to that dual aspiration towards refuge and company. Pascal's pensée in praise of solitude in a room could only be the work of the Auvergnat he was. Alone in his room the Italian feels deprived of all that awaits outside: eyes meeting, bodies brushing up against each other, the endless promise of

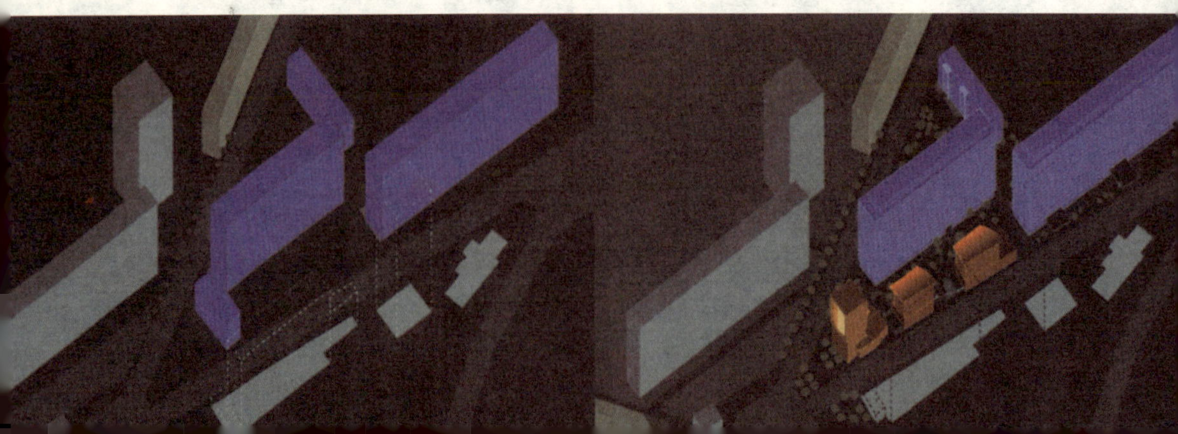

不同建筑的共生，创造空隙，运用深景，调节光线——这些全都是城市对话的新手段。

Achieving coexistence between different buildings, creating interspaces, using vista and light - all part of a new strategy for urban dialogue.

会沉迷于构建某种结构，而会去关注事物发展过程中的相互影响。他们是嬗变之人。与此相悖，怀有乌托邦理想的人无论如何不具备辩证思维，它有什么用武之地呢？这些人憎恨运动与变化，因为这会扰乱秩序，而他们自己却是逻辑主义者。”（吉勒斯·拉布吉，前文引用的文献）

鲍氏回归欧洲、法国和意大利的（传统），伯拉孟特与维杰瓦诺是这一传统的鼻祖。“意大利摒弃的，”鲍氏在多米尼克·费尔南德兹（Dominique Fernandez）的书中读到（多米尼克·费尔南德兹，《拉丁文明》，巴黎，奥利维尔·奥根，1986 年，Dominique Fernandez，Civilisation latine，Paris，Olivier Orban，1986 年），“是空间宽阔、通风良好的法式广场，缘因它们是通行的场所，而非提供庇护、身体恢复或亲密接触的地方。罗马人从不在人民广场（Piazza del Popolo，也称波波洛广场——译者注）驻足，因为那里无法实现自己的美好梦想。他们惟一能做的就是横贯穿行。它与巴黎的协和广场一样典型，都和意大利的神韵气质背道而驰。为什么对围合明确的广场情有独钟呢？因为意大利的广场是生活的场所。在法国，生活起居在自家室内展开。出门仅意味着从一个室内空间过渡到另一个。而在拉丁国家，你会在户外而非家中生活；然而，作为供人生活的场所，户外空间需要与室内空间具有某些类似性质。广场较房屋优越之处在于它能集拢大量人群。拉丁人无法忍受家庭的桎梏，这使他感到断绝了与他人的交际。这是首当其冲之事。广场回应了人们对庇护与交往的双重渴求。帕斯卡盛赞退隐独处的幽静，这只可能是像他那样的奥弗涅人的志趣所在。意大利人在自己房中与世隔绝时，会觉得被剥夺了外界的一切期待：眼神的碰撞，身体的接触与摩擦，机遇与生命的美好展望。固守家中被视为一种谬

chance, of life. Staying home is seen as an error, a sign of selfishness and mean-spiritedness. The square symbolizes the Italian agenda of affectionate solidarity between all members of the community, just as the room symbolizes the French agenda of individual ascesis and perfection through effort."

Portzamparc reflected, hesitated: Place de la Concorde or Piazza del Popolo? Place Vendôme or Piazza Navone?

This was when he began drawing up his "open block" theory: a synthesis in the form of a compromise or, better maybe, a compromise in the form of a synthesis. In 1974 his unbuilt La Roquette project was already pointing to the strategy to come; with Rimbaud, he knew that "inventions by an unknown demand new forms". Working with Georgia Benamo in 1975, Portzamparc designed an apartment complex on the Rue des Hautes Formes, in Paris' 13th arrondissement. Completed in 1979 it featured all the architectural and urbanistic rules that have become his trademark: a landscape subtly broken up into seven buildings linked by arcades, with doors giving onto an open passage and a small, peaceful, light-filled square.

艺术家工作室上方的文化中心，既是标志性讯号，也塑造了（建筑的）整体形态。

The cultural center set above the artists' studios functions simultaneously as a signal and a element giving shape to the whole.

1725 VR 60

误，自私与精神萎靡的征候。广场象征了意大利社会成员间团结友爱的精神，恰如（独立）房间见证了法国人苦行脱世的气质与极力追求完美的尽善论取向。"

鲍赞巴克反思犹豫、踯躅徘徊：协和广场还是人民广场？凡登广场（Place Vendôme，也称旺多姆广场——译者注）还是那佛纳广场？

这时他开始建构"开放式街区"理论：折中形式的合成品，也许会好一些——合成形式的折中。1974 年未能建成的拉·若凯特方案已经预示了昭然若揭的策略；正如兰波所言，鲍氏意识到"无名小卒的发明创造需要新颖的形式"。1975 年鲍赞巴克与乔治亚·班纳莫合作在巴黎 13 区欧风路设计了一座公寓综合体。它在 1979 年落成，成为展示鲍氏在建筑和城市领域一切准则与特色的商标。景观被微妙地分解，7 座独立的建筑最终形成，它们由拱廊联系，出口面向一个开放的通道，也是一座小巧、宁静且阳光明媚的广场。一切应有尽有：这里是鲍氏的城市宣言，展示了他对"破碎"的偏好与对细部的关注，表达了他重归基点与根本问题并（对设计）进行再雕琢与加减变更的决心。体量与节奏的交响再次鸣起，序列与开口相映成趣，这些主题贯穿他的整个职业生涯——无论是特定的建筑设计还是总体的都市项目。

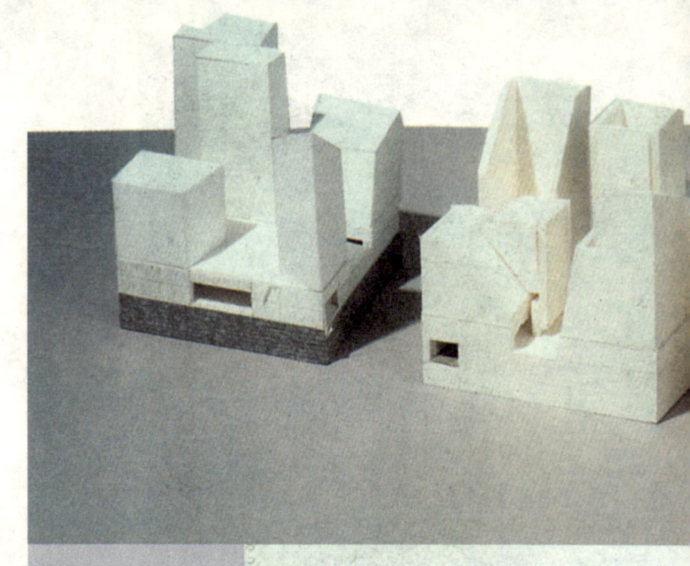

随着研究、探索、建造的进展，鲍赞巴克的"开放式街区"理念逐步精致、日趋完善。
As his studies, research and constructions progressed, so Portzamparc's "open block" philosophy gained in finesse and conviction.

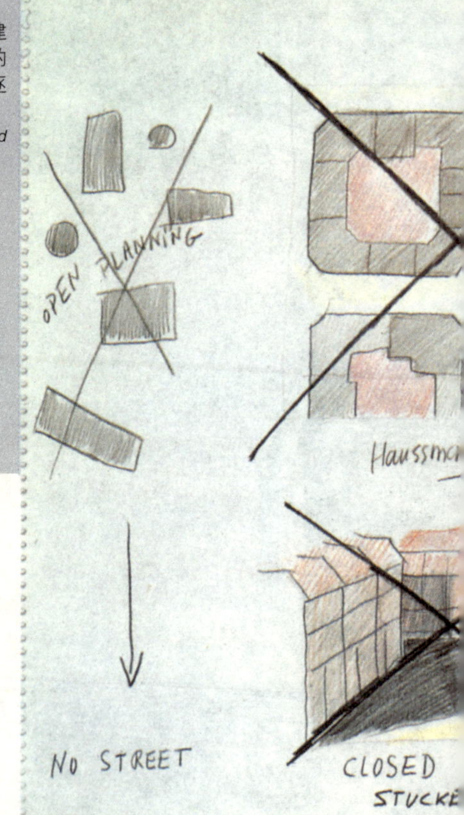

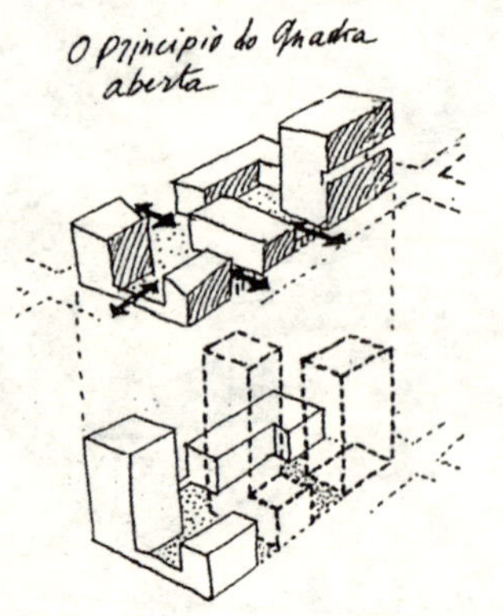

Everything is there: his discourse on the city, taste for fragmentation, attention to detail, and that readiness to rework the original, to resculpt, add and subtract. Here too is the interplay of volume and rhythm, line and opening, to be found from one project to another throughout his career — whether the project be specifically architectural or generically urban.

Creation of relationships and tensions capable of holding dissimilar objects together: such was overriding ambition of a Christian de Portzamparc who outstripped all others in the use of the fragmentation Gilles Deleuze saw as the emergence of "blocks of sensation".

"The idea of fragmentation," said the architect "is, instead of making a finished object, to divide it up into bits. I often use it to create sequences and get a respiration going between what is interior, closed and more or less well lit, and what's luminously open to the sky." Fragmentation opens up spaces — interspaces — using the same idea that underlies his work on the city. "I often talk about the city because it's the collective, civilizing goal towards which architecture should be striving.

"The city is something so complex that mere technical thinking can't keep up with it. It's not just transport plus energy plus esthetics plus sociology plus statistics plus ideology. Add all that up and you don't get a city. There's something else, something mysterious. When you come down to it the art of the city is an idea making its way through history."

A city is also a place to move about in, by day and by night; and like his much-loved Godard and Antonioni, Bataille and Artaud, Portzamparc knows how to discover a city, understand it and tame it — in all its dialectical reality and literary density, naturally. In this he reminds us of author Pierre Mac Orlan, apparently so different in his thinking yet an eloquent advocate of the square: "The Place Clichy reflects a thousand picturesque human details, but it is the Place Clichy itself that reveals them and gives them that literary force we call life. Literature is knowledge of life. A house only has true existence in space

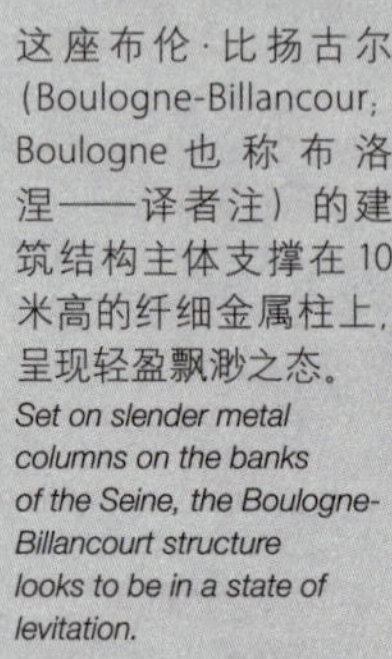

这座布伦·比扬古尔(Boulogne-Billancour; Boulogne 也称布洛涅——译者注）的建筑结构主体支撑在10米高的纤细金属柱上，呈现轻盈飘渺之态。

Set on slender metal columns on the banks of the Seine, the Boulogne-Billancourt structure looks to be in a state of levitation.

四座建筑与视角的变换交响，辅以透明性与浓郁的环境氛围：开放式街区的精华之作。

Four buildings and the interplay of views, plus transparency and atmosphere: the quintessence of the open block.

创造相互间的联系与张力，将相异的个体集结起来：这是鲍赞巴克的最大野心，他在使用破碎理念方面鹤立鸡群，吉勒斯·德勒兹尊称其为塑造“知觉的街区”(block of sensation)。

“破碎理念”，建筑师坦言，“与建造完整事物的希求相反，目标是化整为零。我常常使用它塑造序列，在内部且封闭的空间与一定程度上较明亮，甚或光彩熠熠的空间之间创造此起彼伏的类似于呼吸作用的效果。”破碎将空间展开——交互的空间——他在都市项目中践行同样的理念。“我时常探讨城市，因为它是建筑孜孜以求的共性、儒雅的目标。”

“城市纷繁复杂，单纯的技术思维无法与之匹配。它绝不仅仅是交通、能源、

when it compels acceptance of the amount of literature it contains. The literary impact of an apartment building is comparable, in many cases, to that of a person. Place Clichy is an accumulation of houses, men, women, trams, taxis and sentimental little details ranging from crime to the drabbest honesty. You don't need to reduce these fractions to their common denominator so as to add them up, make a book out of them. Literature does the sums for you..." (Pierre Mac Orlan, preface to Alfred Döblin, *Berlin Alexanderplatz*, Paris, Gallimard, 1970)

The Hautes Formes project is a splendid résumé that reveals much of what inspired Portzamparc and drew his admiration: Wright and Mondrian (whose organization of space began with or moved toward the infinite), Michaux and Stravinsky (who knew what modernity was all about), and Cendrars and Marinetti (who

美学、社会学、统计学、意识形态的结合体。将这一切叠加起来，你也不足以得到城市。还存在其他构成元素，神秘的事物。你细细揣摩就会发现，城市艺术是从沧桑历史中徐徐踱来的奇思妙想。"

无论白昼与夜晚，人们都会漫步流连于城市之中；鲍氏十分赞赏戈达尔、安东尼奥尼，巴塔耶，阿尔多等人，也如他们一样具有发现城市、理解并驯服它的能力——自然而然地应对城市蕴涵的种种辩证现实和奇高的"文学"密度。我们由此联想到皮埃尔·麦克·奥兰，尽管思维方式迥异，但他同样是"广场"的忠实倡议者："克里西广场映射出千姿百态的人文细节，然而揭示并赋予其文学化生命力量的终究是克里西广场本身。文学是关于生命的知识。一座建筑只有承托秉持它应囊括的文学意义时才能真正存在于空间中。很多情形下，一座公寓的文学影响可与一个人相类比。克里西广场是男人、女人、有轨电车、出租车，及其他情感细节要素（从罪恶到最固执守旧的诚实）的累积。你无需将这些片断约减至共有特征然后再相加求和，并以此著书立传，文学已然做出了概括与总结……"（皮埃尔·麦克·奥兰，为艾尔弗雷德·德布林所作的序言，《柏林亚历山大广场》，巴黎，格利玛出版社，1970 年，Pierre Mac Orlan，preface to Alfred Döblin，Berlin Alexanderplatz，Paris，Gallimard，1970 年）

欧风路项目是鲍赞巴克辉煌的名片，展露出激励他前行、为他所敬慕的种种人物与理念：赖特和蒙德里安（他们的空间组织肇始并倾向于无限），米修与斯特拉文斯基（他们对现代性耳熟能详），桑德拉与马里内蒂（他们在速度中发掘美的真谛）。鲍赞巴克注定会在 13 区建功立业，这要部分归功于副市长雅克·特布和巴黎房地产委员会主席米歇尔·隆巴蒂尼的支持。国家大道现在已成为城市复兴、城市

discovered beauty in speed).

Portzamparc was destined to find a lot of work in the 13th arrondissement, partly thanks to the backing of Deputy Mayor Jacques Toubon and of Michel Lombardini, head of the City of Paris Real Estate Commission. The Rue Nationale is now home to a masterly venture mixing rehabilitation, urban recomposition, new housing blocks, gardens and a cultural center: in short, a neighborhood reinvented. There was, too, his management of the Masséna sector, as chief architect: "There's a terrible element of risk involved: a building is one thing, but a whole piece of a city is another." Here he had to invent a scenario, write the rulebook, coordinate the work of other architects (a list as diverse as Chaix and Morel, Norman Foster, Gaëlle Peneau, François Chochon, Olivier Michelin, Rudy Ricciotti), and reconcile styles, aims and a proliferation of materials and colors. It was back to music again, but this time as leader of the orchestra.

In 1996 he won a competition in Boulogne, the outcome being a fresh open-plan block now exclusively occupied by the Canal+ television company. Facing the site was the island of Île Saint-Germain, whose mass of foliage seemed to be floating on the Seine as Dubuffet's tower-sculpture and various factory chimneys reached skyward. Portzamparc divided his plan into four blocks playing on perspective, transparency, atmosphere and a whole range of exterior views. Set 10 meters above ground on slim metal posts, the blocks seem to be levitating; but the impression of ethereality is countered by the use of assertive textured concrete in unusual green-ocher and green-gray shades that dialogue with the sky, the Seine and the foliage.

The Con Edison project in New York offered a radically different challenge, but the response, dictated by the city itself, simply consisted of replacing the horizontally open-plan block with a vertically open version: openings providing extraordinary views, set-back buildings, crossing points, and more. There was, too, the determination to follow the streets down to the

娴熟地控制比例，落落大方地运用带纹理的混凝土。
Beautiful handling of proportion, using a handsome textured concrete.

身为位于巴黎 13 区的马塞纳区管理部门的首席建筑师，鲍赞巴克涉猎的领域不会仅桎梏于建筑学，他还会眷顾城市的一整片区域。
As head architect of the Masséna sector in Paris' 13th arrondissement, Portzamparc's sights were set not on architecture alone, but on a entire section of a city...

……于是他制定相应途径，编制导则，协调其他建筑师的工作。从左至右：弗朗索瓦，尼古拉斯·米歇尔林，查克斯和莫雷尔。(François Chochon, Gaëlle Penault, Nicolas Michelin, Chaix et Morel).
...So he draws up a scenario, writes a rulebook and coordinates other architects. From left to right: François Chochon, Gaëlle Penault, Nicolas Michelin, Chaix et Morel.

重构、新建街区、庭园，和一座文化中心的饕餮盛宴：简而言之，邻里的再发现。还应关注的是，鲍氏曾成为马塞纳区的首席建筑师，负责规划城市区域的发展："这项任务风险重重：建筑是一回事，整整一片城市则另当别论。"他必须拓展特定途径，制定导则，协调其他建筑师的工作（一份风格多样的名单：查克斯和莫雷尔，诺曼·福斯特，弗朗索瓦·奥利维尔·米歇尔林，鲁迪·利奇奥迪），调和各种风格、目标、材料与色彩的扩展。他重入音乐之门，但这一次的身份是指挥者。

1996 年，鲍氏在布伦赢得了一项竞赛，成果是一处具有开放平面的街区，现今被 Canal+ 电视公司全权占有。与基地遥相呼应的是圣 - 日尔曼的岛屿，岛上郁郁葱葱的密林如浮萍般飘荡在塞纳河上，恰似艺术家迪比费的塔式雕塑，又像工厂中密密匝匝、直冲云霄的烟囱。鲍赞巴克将平面分割为四部分，游弋于透视效果、透明性、环境氛围等主题中，构建起一系列完整的室外视角。建筑体块支撑在 10 米高的纤细金属柱上，身形轻盈飘渺；与此同时，带纹理的混凝土不落俗套，呈现灰绿或绿底赭石色阴影，与天穹、塞纳河、密林交相辉映，为建筑飘逸的身影平添几分厚重。

纽约爱迪生电力公司项目则提出了迥然相异的挑战，城市自身做出了回应，水

平方向的开放平面被竖直方向的开放性所取代:豁口提供了非凡的视角,逐层缩进的体量,和(多个造型)交叉点。建筑试图创造沿街道而下,直达东河之感,同时在整组建筑的核心设置一座格拉梅西公园式的庭园。鲍氏竭尽心力展现的是随机性与秩序感的谐和,乌托邦理想与四射激情的结纽,现实与欲求的共存,结构性与辩证主义的博弈。

生于卡萨布兰卡的鲍赞巴克,时常会思忖常驻摩洛哥的休伯特·利奥泰将军的成就,他曾致力于各类宫殿、城墙、清真寺、土式浴室、后宫、fonduk的整修与复原;主持修建港口、道路、铁路、堤坝与发电站等设施。利奥泰身兼任开发商、规划师双重职位,还是身手不凡的建造者,他在斐兹、马拉喀什和拉巴特城外"创造"了三座新城,除这几项壮举,卡萨布兰卡的安法区也值得敬慕,它将古典主义与现代性融会贯通,综合运用阿拉伯-安达卢西亚,装饰派与现代主义的风格。出于政治原因,1925年利奥泰的职位由贝当接任;在登上离开卡萨布兰卡返回法国的航船之际,他说道,"最糟糕的事情是我永远无法建造另一座城市了。"

East River and provide the heart of the block with a garden reminiscent of Gramercy Park. In all these cases Portzamparc willingly took on a grueling brief that involved a rapprochement of chance and orderliness, Utopia and passion, the real and the desirable, structure and dialectic.

Maybe Portzamparc, born in Casablanca, sometimes finds himself thinking of the achievement of Moroccan Resident-General Hubert Lyautey, who was responsible for: the restoration and rehabilitation of ramparts and palaces, mosques and hammams, harems and fonduks; the building of ports, roads and railroad facilities, and the construction of dams and power stations. At once developer and town planner, Lyautey also turned out to be no mean builder, "inventing" three new towns outside Fez, Marrakech and Rabat. To these three successes can be added the Anfa neighborhood in Casablanca, a marvelous combination of classicism and modernity in a style mixing the Arab-Andalusian, Art Deco and modernism. In 1925 Lyautey was replaced by Pétain, for political reasons; and off Casablanca, on the bridge of the ship taking him back to France, he was heard to say, "The worst of it is that I'll never build another city."

纽约爱迪生电力公司(Con Edison):鲍赞巴克将开放街区理念由水平方向转换到竖直方向。

The Con Edison project in New York: Portzamparc transposes his open block theory from horizontal to vertical.

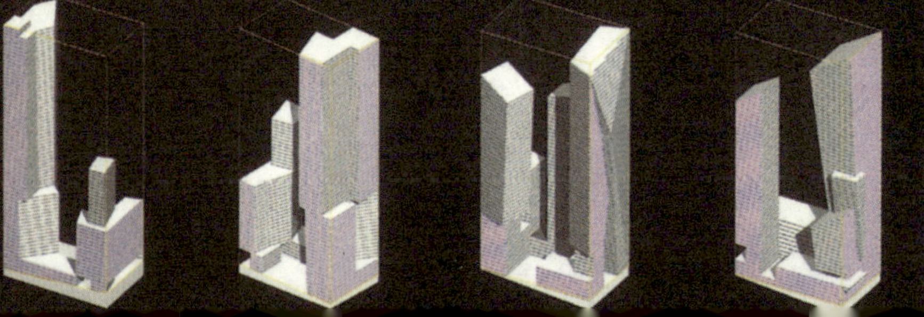

5 升空
taking off

1974 年的"绿色之塔"——位于马恩拉瓦莱的一座水塔——鲍赞巴克的处女作。
In 1974 the "Green Tower"- a water tower at Marne-la-Vallée - is Christian de Portzamparc's first building.

通天塔之谶
The Babel Myth

最初是巴比伦的通天塔。在那里，诺亚的后裔们开始着手建造："他们齐心协力，相互鼓励，共同建设一座城市，一座挺拔高俊、遥可及天堂的通天之塔"（圣经 XI，4）。

他们的目标是：达到神的界域，与之匹敌，并超越之。然而上帝要惩罚这骄傲自负的野心，于是制造出语言的混乱。这是否意味着已经无法相互沟通的人类真正意识到自己妄想与上帝并驾齐驱的努力是徒劳无益的？并非如此：他们认为上帝以自身为原型创造了人，因此人类是上帝的化身。这诡辩深深刺痛伤害到人性，但却也是它最强有力的驱动力来源之一：正如画家罗伯特·马拉瓦尔在 1974 年声称的，"试图掌控一切会使你的头脑混沌不清。我乐于默想业已逝去的亿万岁月，一切的无穷无尽与虚无缥渺，它们如同斯特劳斯的华尔兹和香槟酒一般动人心魄。"人类从未停止对天际的刺探，他们有时如伊卡洛斯一般震动双翅，有时又将双脚坚实地踏在一座座高塔之上：这是宏伟的梦想，力量与崇高精神的展示，绝对与虚无

In the beginning was Babel, in Babylon. There the sons of Noah began building: "And they said to one another, Go to, let us build us a city and a tower, whose top may reach unto heaven" (Gen. XI, 4).

The aim, then: to attain, equal, outstrip the divine. But God in his anger brought low their overweening ambition and created the confusion of languages. Did this mean that men, no longer understanding each other, understood that there was no point in trying to equal God? Absolutely not: if God had created man in His own image, they thought, then man is the equal of God. A sophism that has cost humanity dearly, yet which has remained the source of one of its most potent driving forces: as the painter Robert Malaval put it in 1974, "Wanting to grasp everything makes your head swim so. I like to think about billions of years passing, of infinity and nothingness — all those things as dizzying as Strauss waltzes (and champagne)." Man has never abandoned his thrust towards the sky, wings beating like those of Icarus or, most often, feet solidly set on one kind of tower or another: dreams of grandeur, of power, of elevation, of the absolute, of stardust. Dreams and more dreams.

Babel, though, was just a dress rehearsal. Since then we've seen a succession of launch pads

的桃花源。梦想，弥笃繁复的梦想。而通天塔只是一次彩排而已。自那时以降，我们见证了一系列的“发射塔”以独有的方式出现，远比卡纳维拉尔角的发射塔振奋人心得多：它们的英姿镶嵌点缀着历史长河——从尼罗河谷的金字塔到尤卡坦的金字塔，从拉贾斯坦邦的天文台到芝加哥卢普商业中心——暴风骤雨般浓烈的竞争意识造就了非同寻常的成果，遗产与战利品被一代代传承下去，恰似芭蕾舞剧一般，从纽约跳到上海，从芝加哥跃至吉隆坡。

重温“铸铁设计天才”路易斯·沙利文 1901 年对新建筑秩序的定义将是很有趣的事情：“一座从头至脚，所有细部都已剔除了不谐和线条的建筑拔地而起，它将是浑然一体、崇高和令人敬慕的奇迹。”

整整 100 年以后，当“9·11”已成为往事，我们是否还能以同样的方式认识摩天楼？

通天塔灾难降临后，返璞归真的时候到来了。统一与和谐的神化再次破灭。在《读·写》(Paris，Calmann-Lévy，2003）中，鲍赞巴克告诉菲利普·索莱尔：“我最近曾重游纽约，和贝聿铭一起拜访了工程师莱斯利·罗伯逊，他参与了双子塔的设计。我们共度了一个下午和一个早晨的时光。双子塔是如此抽象炫目，以至于它们与城市已毫无瓜葛。你无法道出 400 米高与 600 米或 800 米的区别所在。当你超越特定高度，摩天楼对环境的影响就成为常量——那是不可测量的——尽管天际线应该另当别论。双子塔是一项壮举，它改变了曼哈顿，然而它却仿佛馅饼一般从天而降。建筑师并未渔利太多。他们婉拒了实录纪事，时至今日，人们仍然无法总结窥探设计建造过程中的种种努力与付出。无可否认，罗伯逊被双子塔的覆灭深深震撼了。他带领我们去了世贸大厦遗址，那里

地域，景观，地标，结构，形象，破碎，包封，折叠，弯曲，嵌入，密实与中空，音乐性：建筑师的设计语汇与特色在这幢处女作中应有尽有。
Territory, landscape, landmark, structure, look, fragmentation, envelope, scrolling, bending, insets, fullness/emptiness, musicality: every feature of the architect's vocabulary is present in this first work.

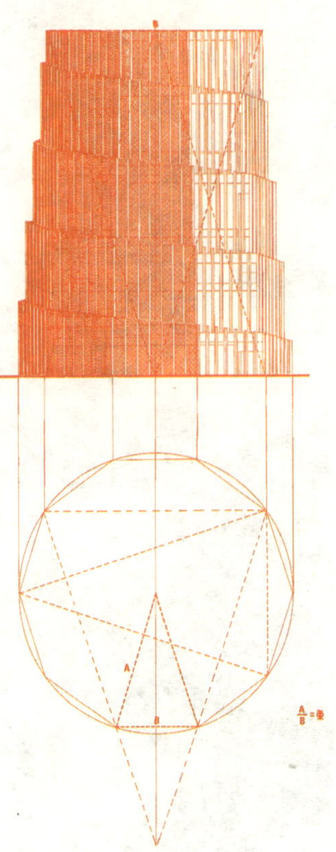

that, in their own way, are more awe-inspiring than Cape Canaveral: history is studded with them — from the pyramids of the Nile Valley to those of Yucatan, from the observatories of Rajasthan to the Chicago Loop — and a fierce spirit of competition has resulted in extraordinary achievements, with the trophy handed on in an endless ballet from New York to Shanghai, from Chicago to Kuala Lumpur.

It's interesting to recall the definition of the new architectural order provided in 1901 by the "cast iron genius" Louis Sullivan: "It must be, down to its tiniest part, a proud and haughty thing, that seems to rise to the heights through the simple marvel of being one, from top to bottom, without a single dissonant line."

Exactly a hundred years later, and with 9.11 behind us, can we still look at towers in the same way?

The innocence regained after the disaster of Babel has flown. Once again the myth of unity has been wiped out. In Voir-Écrire (Paris, Calmann-Lévy, 2003), we find Christian de Portzamparc telling Philippe Sollers: "I was back in New York recently and Pei and I went to see Leslie Robertson, the engineer who had built the twin towers. We spent an afternoon, then a morning with him. The towers were such abstract, dizzying things that they no longer had any scale in relation to the city. You couldn't tell anymore if 400 meters was very different from 600 or 800 meters. Once you go above a certain height the impact within the environment is the same — and immeasurable — even if the skyline is something else again. The towers were real feats that had transformed Manhattan, but it was as if they had been dropped from the sky. Architects didn't take much interest. They defied commentary and even now there's no way of summing up the work that went into designing them and putting them up. Obviously Robertson was deeply affected by the destruction of the towers. He took us to Ground Zero, which was still smoking. The concrete and glass exploded and fused and now you've got earth again; you could grow radishes there."

仍然飘荡着屡屡青烟。混凝土与玻璃分崩离析，熔融流动，并复归于泥土；你甚至可以在那里栽花种草。"

"建筑的所有材料都已化为尘土。基地跌落到地面以下 25 米。这一区域真是动人心魄。这在以前并不为人所熟知，因为双子塔耸入云霄：你无法把握它们，它们是自成一脉的实体，活生生地见证了曼哈顿富有象征性的庄严与宏伟。有人在薄暮时分从布鲁克林拍摄了一幅照片，双子塔映衬着绯红的天空，夕阳穿梭于层层楼板之间，整个建筑变得晶莹剔透；当这两杆岿然直立的魔杖在心中烙下如此抽象的记忆时，你会发觉自己已然望穿了两座中空且又光彩熠熠的形体。罗伯逊告诉我们，他和建筑师雅马萨奇曾经计算过一架 B52 轰炸机对双塔的冲击作用——因为 1948 年一架 B52 曾撞击帝国大厦，那座建筑没有像人们想像的那样被（撞击）摧毁，而是被修复一新。飞机有可能碰撞建筑，因此进行影响计算也顺理成章。他向我们展示了撞击功率，包括飞机的重量与速度；但"9·11"袭击中撞击第一座塔楼的破坏力是 B52 的 10 倍。不要忘记还有满载的燃料：这意味着 20 余倍的爆炸力和高温。"

"对我来说，袭击事件几乎带有某种古旧的中世纪意味：霎那间，在舞台中央，

鲍赞巴克的摩天楼一贯显现出开放的形态，修长纤细，高耸伫立，成为鲍氏的座右铭，也表达了他的志趣：欧风路的中世纪气韵和柯布风格（左），福冈（Fukuoka）的古典主义和异国情调。

Always open, extraordinarily slender and utterly vertical, Portzamparc's towers are a summary of his vision and his architectural virtuosity: medieval and/or Corbusian at Hautes Formes (left), classical and/or exotic at Fukuoka (right).

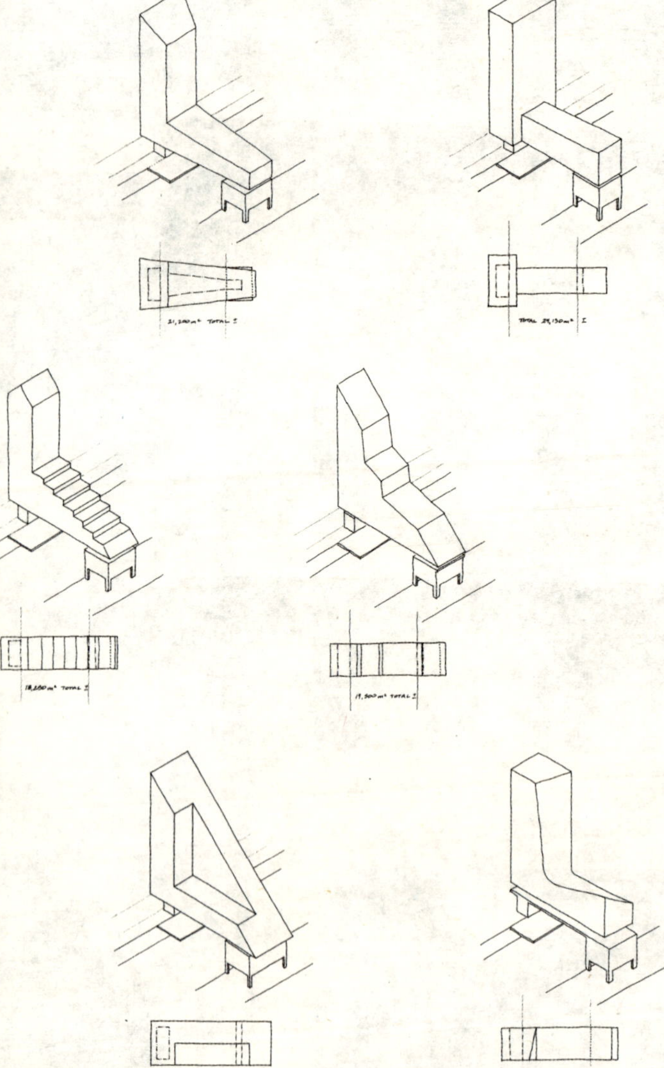

"All the material of the buildings has become earth. And the level has dropped to 25 meters below ground. The area of the site is really striking. This wasn't readily apparent before, because the towers were so high: you couldn't get a handle on them, they were just there as an entity, a living thing existing as evidence of the symbolic grandeur of Manhattan. There's a photo taken from Brooklyn at dusk, where you see the towers totally transparent against a red sky, with the sun filtering through at each story; once you had this abstract impression of two upright rods, then you find yourself looking through two hollow, luminous bodies. Robertson told us that he and Yamasaki, the architect, had calculated the impact a B52 would have on the towers — because in 1948 a B52 had brushed the Empire State Building, chipped it you might say, and it had been repaired. A plane can hit a tower, so you make your impact calculations. He showed us the power ratios, including the speed and weight of the plane; but on 9.11 the impact of the plane that hit the first tower was ten times that of a B52. Then

一座建筑/桥梁70米宽，100米高，位于"欧洲里尔"商务中心区的里昂纳信贷银行大厦横跨里尔-欧洲（TGV）高速列车站。

A building/bridge 70 meters wide and 100 meters high, the Crédit Lyonnais tower at Euralille bestrides the Lille-Europe TGV station.

高度密集和不规则的结构造型阐释了自身的重量与密度。
A highly compact, atypically-shaped structure asserts its weight and density.

悠长的办公区通行走廊，门窗一例面向里尔旧城。
The long internal traffic areas for offices all giving onto the Old Town in Lille.

建筑师的象征性力量与弩炮飞弹直面相击。你体会到远古的传奇在当代上演。"

"圣吉米亚诺（San Gimignano）和锡耶纳（Siena）曾深处枪林弹雨之中，城中的高塔屡屡被攻击，借以威吓其臣民……"

在鲍赞巴克的作品中，摩天楼的象征性意义变化多端，其类型涉及通天塔风格（马恩拉瓦莱的"绿色之塔"），城堡要塞风格（欧风路项目，里尔的里昂纳信贷银行大厦），图腾柱风格（"万代"大厦和路易威登大厦），"坦比哀多"（tempietto）风格（福冈），和耸立之石（赫斯特大厦），北方大厦（Septentrion），T1 大厦，磐石大厦）。

there were the full fuel tanks: that meant an explosive force and temperatures twenty times higher.

"For me there was something almost archaic and medieval in that attack: all of a sudden, center stage, you've got the symbolic power of the architect facing that of the cannonball fired from a catapult. You have the impression of a very ancient story being played out.

"San Gimignano or Siena was once the target of enormous cannonballs, fired in at their towers to make the people afraid..."

In Portzamparc's work the symbolic potential of the tower is astonishingly varied, the range covering the Tower of Babel (the "Green Tower" at Marne-la-Vallée), the castle keep (the Hautes Formes, the Crédit Lyonnais Tower in Lille),

这一切——或几乎是一切——在1974年的马恩拉瓦莱项目中被表达、书写、建造并确认。通过对建筑师基本设计语汇和标志的概括——地域，景观，地标，结构，造型，破碎，包封，折叠，退台形式，密实与中空，封闭与开敞，音乐性等等——它揭示了某种发展轨迹与许诺，成为绘声绘色的宣言，使我们得以重温通天塔，顺带也重游了设计师塔特林的世界。

"当我着手设计这座30米高的建筑时，我曾扪心自问怎样才能使它看上去有40米或60米高。我使用了螺旋形态，螺纹的间距是7米——与周边计划修建的公寓和办公楼尺度完全不同。盘旋上升的倾斜与动感，使其傲视周边的地标建筑，赋

the totem pole (Bandai, LVMH), the tempietto (Fukuoka) and the standing stone (Hearst, Septentrion, T1, Granite).

Everything — or just about — was said, written, built and affirmed at Marne-la-Vallée in 1974. Babel revisited and Tatlin mentioned in passing, you might say, in a tangible manifesto that would establish both a trajectory and a commitment via its summary of the architect's basic vocabulary and stamp: territory, landscape, landmarks, structure, figure, fragmentation, envelopment, folding, set-back shape, solid and hollow, closed and open, musicality — and the list goes on.

"When I was putting up this 30-meter building, I had asked myself what could be done to make it look 40 or 60 meters. I used a spiral with a basic thread of seven meters — a measurement

东京的"万代"大厦囊括剧院，电影院，展览，居住设施和办公等各项功能，希图象征信息通讯的世界。

Containing theaters, movie houses, exhibition spaces, apartments and offices, the Bandai tower in Tokyo was intended to symbolize the world of communications.

立面永远笼罩在变幻莫测的色彩中，为东京美轮美奂的电气世界歌功颂德。

Constantly swept by changing colors, the facade was a tribute to the electrical beauty of Tokyo.

纽约路易威登大厦的构思草图。
Preparatory drawings for the LVMH tower in New York.

弯曲而成的棱柱形体高 112 米。不规则的白色磨砂玻璃立面，产生出光洁如丝绸一般的颤动感和冰清玉洁的衍射效果。
Bent into the shape of a prism 112 meters high, with irregularly aligned, sandblasted glass areas on the facade, the building sets up silken vibrations and snow-like diffractions.

予它纪念碑式的尺度与风采。我将螺旋体表皮的构思结合实际的水塔功能，采用半透明的面层嵌板。其效果与其说是墙体，倒不如讲更像是生长有植被的透明格栅。”（因建筑沿上升的螺旋线种植有植物——译者注）

“当你步入建筑之中，空气中弥漫着穿透格栅倾斜而下的阳光，散发着迷人的气息。而从建筑外侧可以看到水箱和混凝土壳体背光而立的身影。这水塔不是某一事件、某位名人的纪念碑；真正的意义在于它和景观肌理的体量关系——它自身的纪念性。”

鲍赞巴克在与皮埃尔·布莱兹相遇之前 10 年就开始执迷于螺旋体。有趣的是，在谈到自己的作品《应答曲》时，布莱兹曾提到螺旋体是同时具备开放性（它可以无限延伸）、封闭性与整体性的形体（它可以在任意点被截断，形式仍然保持完美，绝无瑕疵）。

在欧风路住宅中，我们见证了垂直

totally different from that of the apartment and office blocks scheduled to go up round about. Combined with the obliqueness of the spiral movement, this sets the structure apart from the surrounding landmarks and gives it monumental scale and presence. I applied the spiral envelope idea to the actual water tank as well, using semi-transparent panels. The result is walls that aren't walls, but rather a transparent grille with vegetation growing on it.

"When you're inside, you have a surprising impression of light filtering through the grille. From the outside you see the tank, the concrete shell, lit from behind by the sun. The water tower is not a monument to this or that; what counts is its size relationship with the landscape — its monumentality."

So Portzamparc was into spirals ten years before meeting Pierre Boulez. But interestingly, speaking of his composition Repons, Boulez also mentions the spiral as a shape simultaneously open (it can be extended infinitely) and closed and entire (it can be halted at any point and its form remains perfect, lacking nothing).

With Hautes Formes, even if we're looking

版本的萨伏伊别墅——对柯布西耶的颂词——但同时对雉堞堡垒的参照也是显而易见的；在福冈，则是一座独立形体的高塔，顶部设有避风遮阳的平台，它恰似坦比哀多礼拜堂一般，还使人联想到日本武士和幕府时代的将军。

"城堡"风格在 1995 年建造的里昂纳信贷银行大厦中再露端倪。这座要塞形如建筑和桥梁的结合体，70 米宽、100 米高，是设计与技术的集大成之作，拥有不规则的形态：极度紧凑的体量彰显了自身的厚重与高密度，它坐北朝南，向城市散发着熠熠光彩，仿佛巨大的字母 L，在基座之上飘浮摇曳，并不停滞呆板。

这种充满辩证气质的对场地进行系统

建筑顶部12米高的巨型大厅被玻璃墙面围合，与摺拢在立面上犄角旮旯的霓虹灯一起编织了一首色彩与形式的交响曲。

An immense glass-walled hall 12 meters high tops the LVMH tower, with neon lighting set into the nooks and crannies of the facade to generate a symphony of color and form.

规划的第二座路易威登大厦强调倾斜交错，原本可以和第一座塔楼默契相合。

The planned second LVMH tower, with its emphasis on obliques, would have gone marvelously well with the first.

at a verticalized Villa Savoye — and thus at a tribute to Le Corbusier — the reference to the castellated keep is clear; at Fukuoka an isolated shape topped with a terrace sheltered from wind and sun, speaks of the tempietto at the same time as it conjures up samurai and shoguns.

The castle keep returned with the Crédit Lyonnais tower that went up in Lille in 1995. Taking the form of a combined building/bridge 70 meters wide and 100 meters high, the keep is a planning and technological feat resulting in an atypical shape: an extremely compact mass that

性、因地制宜处理——并非融为一体——的设计模式在鲍氏的两座"图腾柱"式摩天楼中得到更深层次的表达，虽然两座建筑迥然相异。

在东京，他在 1994 年赢得了"万代"大厦的设计竞赛，这座巨无霸式的建筑位于城市心脏地带的青山道地区，旨在象征交流与信息通信，它囊括剧院，电影院，展览，居住设施和办公等各项功能。对鲍赞巴克来讲，东京迈跨着"混沌踉跄的步伐"，丧失了一切参照点与秩序，使他得以重归罗兰·巴特的世界："整个城市凝聚在一处封闭、漠然的场所，伪装于城壕与草木的面具之内，有一位从未谋面的君主蛰居于此——换言之，一位未知者。"重温罗兰·巴特，鲍氏发现自己惊诧于《银翼杀手》(Blade Runner）和电子游戏的交响曲，光在这里成为主题：（他据此设计了）永恒且变化多端的发光立面，为东京美轮美奂的电气世界歌功颂德。

"我对照明与光的构思起始于一个清晰明确的主题：它延续拓展了我为巴黎音乐城大音乐厅设计的巨大声学展廊的构思。我设置了一系列电脑控制的发光管，这意味着（建筑）能够发出任意波长的光线。"

事事无常终难料，资金并不总能及时到位，这脉动的、变幻的光的溪流最终未能成形。

与东京相较，纽约则另当别论。

"你常须谨记的是，"路易斯·费迪南德·塞利纳 1932 年曾在《暗夜旅程》(《Journey to the End of the Night》) 中写道，"他们的城市是垂直的——完完全全的垂直。纽约是站立的城市。我游历过很多美丽的城市和港口。它们沿海岸线或河滩平展开来，在景观基质中铺伸绵延，静候旅行者的到来。但他们中没有一座能与美利坚版本相媲美，纽约直立起身体，坚挺无

北方大厦的方案强调棱柱体形的倾斜与垂直，此类构思曾在纽约的路易威登大厦项目中牛刀小试。

Portzamparc's Septentrion project stresses the obliques and verticals of the prismatic approach so successfully begun with the LVMH tower in New York.

makes no secret of its weight and density and yet, flaring out and facing southwards towards the city like an enormous L, appears to float on its base rather than rest on it.

This dialectical play with systematic adaptation — not integration! — to the site would find further expression in two Portzamparc "totem poles", each utterly different from the other.

In Tokyo, in 1994, he won the competition for the Bandai Tower, a monster in Aoyama-Dori in the heart of the city, and intended to symbolize communication while containing theatres, cinemas, exhibitions, living accommodations and offices. For Portzamparc Tokyo is a "chaotic shambles", bereft of reference points and order, that sent him back to Roland Barthes: "The entire city is centered on a place at once forbidden and indifferent, masked with greenery, moated, and inhabited by an emperor one never sees — in other words, by an unknown." After re-reading Barthes he found himself wondering about Blade Runner and a symphony of electronic games, with light as the key: a tribute to Tokyo's electric beauty, with a permanently, ever-changingly lit facade.

"I began with a lighting idea based on a very explicit theme: the idea behind one of the enormous acoustic windows I had designed for the large concert hall at the Cité de la Musique in Paris. I'd put in a set of computer-driven color tubes that meant you could work through the entire spectrum."

Ultimately, however, this stream of moving, changing color never came to be, a frequent turn of events in a field in which the money is not always around when you need it.

New York is a completely different kettle of fish than Tokyo.

"What you have to get into your head," wrote Louis-Ferdinand Céline in Journey to the End of the Night in 1932, "is that their city was upright — absolutely upright. New York is a city standing up. I'd already seen plenty of cities, beautiful ones, and great ports as well. But where I come from, the cities lie down flat along the coast or the banks of rivers, they stretch out along the

比：不似女性般阴柔，却如男子般刚强。"

"不似女性般阴柔……"当人们谈及纽约建筑，常常会被误导，过度执迷于伯纳姆具有革命性的熨斗大厦、凡·海伦的装饰派风格代表作克莱斯勒大厦，赖特的巴洛克风格代表作古根海姆博物馆，密斯·凡·德·罗的极少主义代表作西格拉姆大厦，甚至颇具幽默气质的后现代主义小乘之作，约翰逊的 AT&T 大厦（美国电话电报公司大厦）和它"齐本德尔"式的三角山花。这些建筑使人们忘记了纽约其实是一个繁冗单调、极具重复性的城市，它被一系列森严的政策法规所桎梏。

"不少纽约人询问我到底用什么方法做出了别人未曾做出的设计？我的回答是在建筑的每个自然层都牺牲一部分面积，并征得许可，不必将立面设计得与街道完全平齐。谁都认为这违反了分区法规，但当我咨询是否能将立面退进时，资深顾问回答说：'有什么不可以呢？'"

"然而，许可的退进程度要经过严格的计算。我运用了一个能够平衡取舍的十分简洁而微妙的规律：抛弃某些体量（与空间），另寻它处予以补偿。这是不乏趣味、但严格'商业化'的决断，没有'风格'上的偏见。在巴黎，玩弄立面是天方夜谭：（立面）必须严格对齐，有关悬挑和退进的规定完全出自古典建筑传统。"

"在纽约的这个项目中，我希求打破常规的阶梯式形态，创造强势的直截了当的垂直性。迈克尔·佩利告诉我无须非得像阶梯一样以水平方式实现退进。你只需将建筑沿直线伸展，考虑好各部分间的联系即可。可见在规则手册中也包涵着自由，尽管极少被援引，甚至无人知之：因为在纽约，浪费哪怕一平米的面积，或不把每一楼层塞满，不追求高度与面积的最大化，那是不可想像且缺乏理性的。"

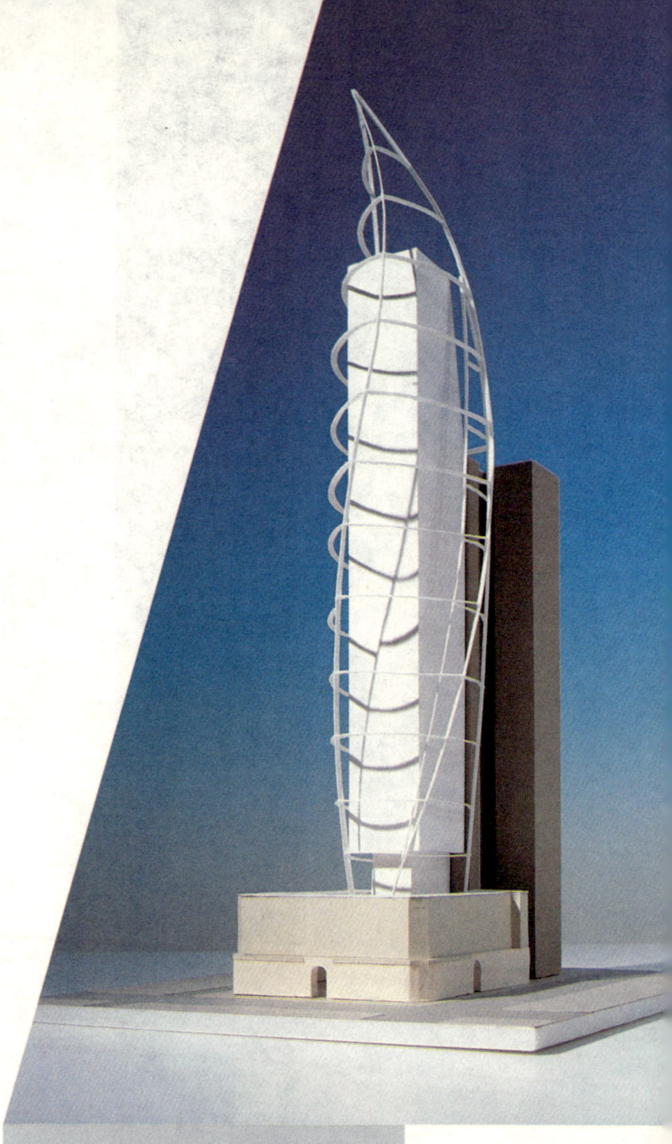

无愧于心的精湛技巧和纯粹的空间几何学：然而，赫斯特大厦（Hearst, 2000）永远无法矗立于纽约的天际线之中。

Unashamed virtuosity and pure spatial geometry: but the Hearst tower will never take its place on the New York skyline.

landscape, waiting for the traveler. None of this swooning for the American version, it stands up straight and hard: not female-fuckable, but scarily male-erect."

"Not female-fuckable..." Where New York buildings are concerned, we're too often misled by the revolutionary splendor of Burnham's Flat Iron, the Art Deco splendor of Van Halen's Chrysler, the baroque splendor of Wright's Guggenheim, the minimalist splendor of Mies Van Der Rohe's Seagram, and the little masterpiece of postmodern humor that is Johnson's AT&T, with its "Chippendale" pediment. They make us forget that basically New York is a monotonous, repetitive city with an apparently draconian set of regulations.

"Lots of New Yorkers asked me, How the hell did you manage it, why weren't other people doing it? My answer was that I began by sacrificing a few square meters of empty space on each floor and asking permission not to have to align the facade with the street. Everybody thought this was contrary to the zoning code, but when I asked if you could set a facade back, the specialist consultant said, 'Why not?'

"However, there was some razor sharp calculating to be done on how much setting back you could get away with. Then I came across a very tricky little regulation based on give and take: give up some volume here, take it back somewhere else. This is a very interesting, strictly 'commercial' ruling, with no preconceptions about 'style'. In Paris, messing round with facades is out of the question: the alignments are strict and the rules on overhangs and setting-back come straight out of classical architecture."

"In the New York project I wanted to break with the typical staircase silhouette and try for powerful, direct verticality. Michael Purly told me, 'You don't have to follow the alignment of the set-back horizontally, like on a staircase. What you have to do is make a building along the line, and a point of contact is all you need. So in fact there's a freedom in the rulebook that's little used and probably little known: because in

这是市场、金融、标准、规范、技术禁锢风格之创造的又一实例。

鲍赞巴克历经四载，精明睿智地摆脱重压，1999 年在纽约天际线中塑造出了一座风格独具的"另类"建筑：坐落在 57 街的路易威登大厦，位于第五大街和麦迪逊大街之间。

建筑高 112 米，像棱柱体一样弯折。白色的磨砂玻璃立面和不规则的线条产生出光洁的、丝绸一般的颤动感和冰清玉洁的衍射效果。它似乎从天而降，满怀丝绸、瑞雪和珍珠般闪亮的彩虹色气宇，带给曼哈顿奇异的音乐风韵，这一切让纽约时报建筑批评家休伯特·马斯卡姆目瞪口呆，他在官方剪彩仪式后称其具有"爵士风格"。

在纽约立足意味着打破本地固有的几何形态，闯入光影与映像的世界——尤其与街对面凝重阴郁的 IBM 大厦相对而立——并建造真正的建筑，而非立面。他成功的秘诀与画龙点睛之笔是塔楼顶部被玻璃围合的 12 米高的大厅，入夜后，这里变成了魔幻般的城市泛光灯，与摺拢在犄角旮旯的霓虹灯一起编织起赏心悦目的视觉合声交响曲。

不久后，路易威登公司购买了 57 街与麦迪逊大街交角处的地块，获得了拆迁许可，由伯纳德·阿诺特出面邀请鲍赞巴克再次出山。关键问题在于避免重复的同时还能取得和谐效果。构思草图在马尼拉至香港的一次航班上产生。第二座摩天楼与第一座迥然相异：例如，它的限高是 80 米；造型采用平行六面体，突出斜线的灵动与交错，使第一座塔楼具有更佳的可视性，同时自身也以优雅的姿态镶嵌入拐角处的基地，柔美、谦逊、谨慎；在它的上部，梯形的棱柱体向内收分，形成独特的造型。

这一切凝聚成一座生龙活虎般的磨砂玻璃体形，主色调为灰色——没有珍珠与白雪的雍容华贵——它如同一尊巨大的浮

为拉·德方斯所作的 T1 大厦，如武士的利剑一般刺入云霄。
The T1 tower planned for La Défense, in Paris, seems to slice through the air like a samurai sword.

New York it's unthinkable, irrational, not to use every available square meter and cram as much as you can into every story, while repeating your square meters by building higher and higher." Another illustration of the way the market, finance, rules, regulations, and technical considerations weigh on matters of style.

Portzamparc spent four years subtly getting out from under the weight and delivering, in 1999, a fresh "exception" to the New York skyline: the LVMH building on 57th Street, between 5th Avenue and Madison.

One hundred twelve meters high, the building is bent like a prism, with the white sandblasted glass of the facade, all irregular lines, generating silken vibrato and snowy diffraction. Out of nowhere the combination of silk, snow and pearly iridescence quickened Manhattan with a strange musicality that much-feared New York Times architecture critic Hubert Muschamp would describe, after the official opening, as "jazzy".

For Portzamparc, existing in New York meant fracturing the local geometry, breaking into the reflections — especially of the somberly overpowering IBM presence across the street — and putting up a real building, not just a facade. The measure of his success is that at night, the enormous, glassed-in room 12 meters high that tops the tower is transmuted into a magical urban luminaire, with neon tubes tucked away in the nooks and crannies and combining to form a stunning visual and chromatic symphony. Not long afterwards, LVMH bought the neighboring lot on the corner of Madison and 57th, got demolition approval and in the person of Bernard Arnault asked Portzamparc for a repeat performance. The big problem, of course, was to avoid running up a twin while ensuring harmony. The idea and the sketch bubbled up in a plane between Manila and Hong Kong. The second tower would be resolutely different from the first: its height, for one thing, would not exceed 80 meters; the parallelepiped would be dropped in favor of an interplay of obliques that would mean far greater visibility for the

拉·德方斯的磐石大厦再次应用几何学理念；但棱柱体的概念让位于在两个二面体之间设置圆柱形体的构思。
Geometry again, with the Granite project at La Défense; here, however, the prism is supplanted by a cylinder set between two dihedra.

雕屏风，在远处即可瞥见。

这两座并肩而立的建筑使我们联想起保罗·莫兰（Paul Morand）1929年从纽约回到法国时写下的感言："摩天大楼！它们中有些是女性的化身，其他的则更像是男人：有些貌如太阳神庙，有些却酷似阿兹特克人（Aztec）的月亮金字塔。"

然而太阳与月亮的轨迹注定无法交汇。就在"9·11"之前一个月，项目被取消了。是谁未卜先知？

鲍赞巴克重拾罗兰·巴特："身体中最为色情撩人的部分是衣饰被悠然缓慢掀开的地方。"

"万代"大厦和路易威登大厦的设计——以及更早的"绿色之塔"——使建筑师有机会试验表皮及遮蔽的设计游戏，"显现与消失"的设计主题，对称与非对称的相互博弈。

这些策略在他接下来的一系列项目中被发展、拓宽、并更新，而变化多端的棱柱形体成为主要塑造对象。

在巴黎的拉·德方斯，不足三年时间（2000-2002年），他主持的3个设计项目大功告成。北方大厦的设计，建筑师强调

first tower, while slotting elegantly into the corner site in a mix of tenderness, modesty and discretion; and lastly there was its shape, that of a trapezoidal prism incurvated on its upper part. It all added up to a vivid block of sandblasted glass, mainly gray — not mother of pearl and snow — presenting as a kind of gigantic screen in relief, visible from quite a way off.

These two faces side by side remind us of Paul Morand, writing in France on his return from New York in 1929: "The skyscrapers! Some of them are women, others are men: some look like temples to the Sun, others like the Aztecs' pyramids of the Moon."

Once again, however, the paths of sun and moon were destined not to cross. Just a month before 9.11, the project was called off. Did somebody have a premonition?

Yet again Portzamparc went back to Roland Barthes: "The most erotic part of a body is where the clothing falls open slightly."

With Bandai and LVMH — and before that with the Tour Verte — the architect had tested the game of masks, of appearance/disappearance, of dissymmetry as opposed to assymmetry.

This was a strategy he would develop, expand and renew in a succession of projects, with primacy given to the prism in all its variations.

At La Défense, in Paris, three projects came to a head in barely three years (2000-02). In

倾斜与垂直的理念以及棱柱形态风格——这一构思曾在纽约的项目中牛刀小试。T1大厦则如武士的利剑一般刺入云霄。而在磐石大厦中——鲍氏赢得了竞赛——棱柱的概念让位于在两个二面体之间设置圆柱形体的构思。对空间几何的深刻研习——如果我们允许自己无忧无虑地进入奇异的

his Septentrion tower, the architect stressed obliques and verticals, together with the prismatic style begun and confirmed in New York. The T1 tower seemed to cut through the air like a samurai sword. And the "Granite Tower" — Portzamparc won this competition — saw the prismatic concerns give way to a cylinder set between two dihedra. This fine exercise in spatial geometry

在法兰克福举行的城市设计竞赛中失利的项目，但它为2003年一座新塔楼的设计带来灵感。

An urban development competition, lost in Frankfurt, nevertheless provided the inspiration for a new tower project in 2003.

潜意识世界游历一番的话——可被理解为针对T1大厦的失败所进行的复仇雪恨：垂直的二面体和统一的棱柱被干净利落地切割，如军刀劈削出的一般。

再次回到纽约，赫斯特大厦（Hearst，2000年）的几何形态刹那间燃烧起涌动的火焰，尽管项目仍旧胎死腹中；派克大街上的卡里米安大厦（Kalimian Tower，2003年）将两座垂直的棱柱并置起来，并在建筑一隅或转角处附加了3个柱体体块。它是技术的里程碑，融汇了杰出的几何学计算，而这些却无损于直觉和感知的自由。同样精湛的技巧可在2001年的另一项目中寻得踪迹，它运用了马克·杜提尔制定的城市设计导则：两座宛若一体的塔楼从中间一分为二。它们好似在水中亭亭玉立——底层设有架空柱——建筑下部是一座富丽堂皇的庭园和一幢由让-米歇尔·维尔莫特重新装修的旧宾馆，建筑仿佛薄如蝉翼的陶瓷瓶或双耳陶罐一般；它是对地中海海岸灵动慰贴的膜拜，因为那里有史以来就是自然与文化的摇篮，"幸福、和平与富饶"的发祥地。

could be taken — if we allow ourselves a lighthearted venture into the mysteries of the subconscious — as revenge for the T1 failure: the vertical dihedron and the unifying prism seem cleanly sliced, as if by a saber.

Back in New York again, geometry abruptly became a vibrant flame with the Hearst Tower project (2000), never to be carried through; and on Park Avenue, with the Kalimian Tower (2003), for which Portzamparc brought together two vertical prisms and three prismatic objects added to the corners and sinuosities. This technical eclat and masterly geometrical calculation have proved no obstacle to intuition and sensual freedom. This same virtuosity can be found in a 2001 project which applied the urban design principes set out by Jean-Marc Dutilleul: two towers resembling a single one split down the middle. Standing in water, so to speak — they are set on pilotis — they rise out of a luxuriant garden and an old hotel building revamped by Jean-Michel Wilmotte like slender ceramic jars or amphorae, a vibrant tribute to those Mediterranean shores which, since the beginning of time, have been the cradle of nature and culture, of "pleasure, peace and opulence".

两座塔楼位于地中海的边缘，底层架空，实际上是被一分为二的整体，使人联想起双耳陶瓷瓶的形态。

Set on pilotis on the edge of the Mediterranean, these two structures are in fact a single tower split down the middle in a shape reminiscent of amphorae.

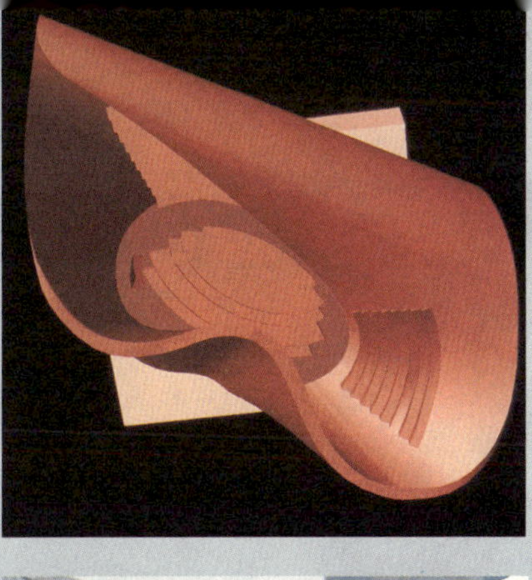

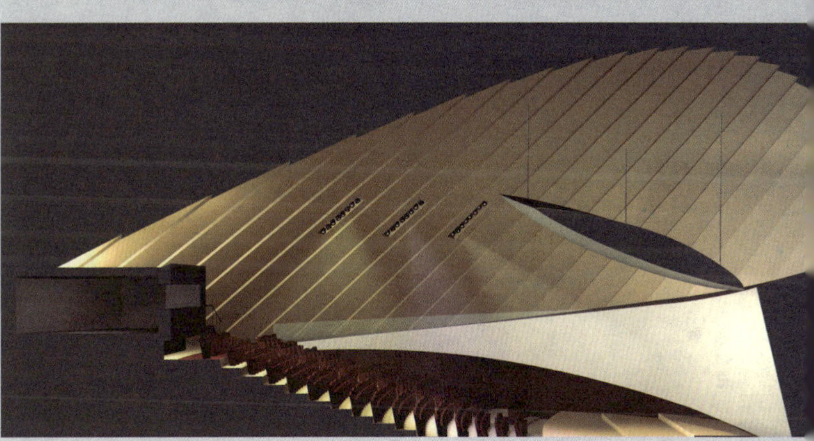

栖居
dwelling

在拉·若凯特项目中，鲍赞巴克在建筑与城市之间建立新的关系。

With his 1974 La Roquette project, Portzamparc made possible a new relationship between architecture and the city .

私人维度

The private dimension

鲍赞巴克的第一次建筑经历可以追溯到什么时候呢？那几乎是12岁的孩童时代，他倏尔突发奇想，决定重新整修卧室，使其沐浴阳光之中、广迎四方来客；又或者建造起隔断，离群索居，如果他喜欢的话。

那时，他摸索着从窗口探出身子，俯视远眺街道的尽头，或者——在更私密的尺度上——审视窗下的庭院：科克托曾说庭院造就了自己的整个世界，而希区柯克却在《后窗》中将其塑造为人类邪恶行径的温床。但似乎对鲍赞巴克而言，与住宅相关的一切都会与童年联系起来。当时被青年们仰慕的偶像——勒·柯布西耶，鲍

How far back does that first architectural memory go? Almost certainly to age twelve, when the child suddenly decides to re-do his room, opening it up to the light and to other people; or, when he feels like it, isolating it and putting up ramparts.

And then he's hanging adventurously out the window looking far down the street. Or — on a more intimate scale — he's checking out the courtyard: even if Cocteau said his courtyard offered him the entire world and Hitchcock in Rear Window made it the very seat of human iniquity. It would seem that for Portzamparc, everything to do with houses has to do with childhood. Despite a list of adolescent models — Le Corbusier, Beaudoin, Candilis — more concerned with "machines for living" than with places to live in, it is fair to say that childhood

顿因，坎迪利斯——往往更加关注"居住的机器"而非宜居的场所，而平心而论，鲍氏则与之相反，"童年"经历一直无所不在。"孩子们在进入梦乡之前，"丹尼斯·蒂利纳克（Denis Tillinac）曾写道，"构思了他们希求的梦境。故事的中间情节虽然各式各样，但结局总是殊途同归：（人们）重回家园，爬上高高的山岗，村庄映入眼帘，田间小径一直通向宅基地，锈迹斑斑的大门（出现了）……而当所有的家宅都贱卖给匿名的异乡客后，我们的一切便都随风而逝了。"（Denis Tillinac，《Maisons de famille》，Paris，Robert Laffont，1987）

这一切毫无疑问充满乡愁与怀旧情感。或者还有少许的复古与保守？然而鲍赞巴克游离于这二者，早达登峰造极之境

remains omnipresent. "Before drifting off to sleep," writes Denis Tillinac, "children construct a dream they hope will see them through the night. Mine had different plots, but the epilogue was invariable: returning home, climbing the plateau, the village coming into sight, the pathway leading to the grounds, the great rusted gate...When all the family houses have been sold off to anonymous outsiders, it will be all over for us."(Denis Tillinac, *Maisons de famille,* Paris, Robert Laffont, 1987)

Nostalgic, certainly. A tad reactionary, maybe? But you couldn't be further from Portzamparc, who's absolutely neither: he's an urban animal who sometimes gets a little carried away. If there's anything provincial about him, it's the provincialism of Jean Renoir's *The Rules of the Game*, or of life at the Villa Noailles at Hyères, in southern France.

And yet this nebulous, mysterious tinge of

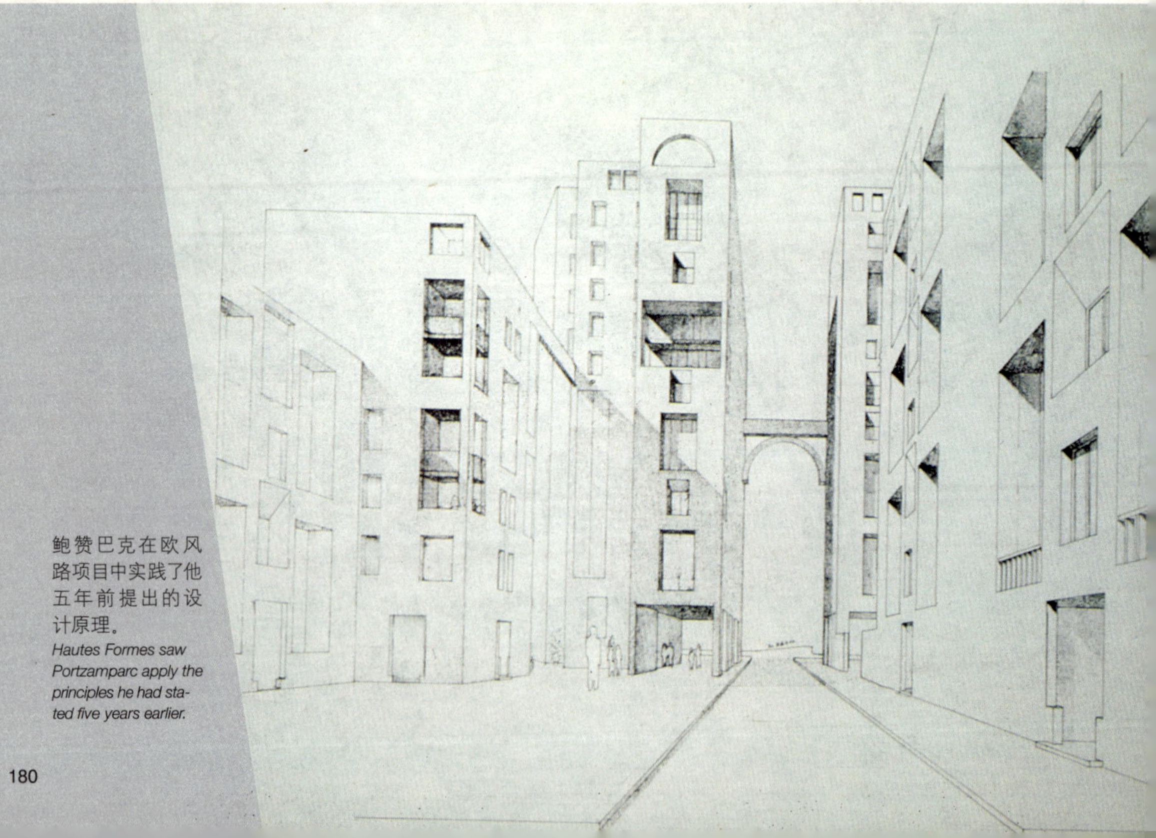

鲍赞巴克在欧风路项目中实践了他五年前提出的设计原理。
Hautes Formes saw Portzamparc apply the principles he had stated five years earlier.

界：他是城市狂人，常怀有一缕神魂颠倒的情思。如果说他有些许乡土观念的话，那便是简·雷诺阿在《游戏规则》中体现出的，或法国南部耶尔城诺埃莉斯别墅生活中显露出的地方风尚。

但这朦胧暧昧、神秘难解的孩童气息弥漫于他所有的住宅项目中：以其特有的方式将户内空间与户外空间合而为一，将室内与室外敷以同等的重要性，展示了梦境与现实的共存。他再次创造着景观，视野，远距离透视效果和飘逸的线条。

"我期待在一座建筑中运用多种元素，"他解释道，"这导致了对冲突与二元性的偏爱。柔顺的美被其他地方的刚直与坚固所揭示。我喜欢创造对比，互补，和感知的二元性……我喜欢塑造浩大而具有扩张性（的效果），但同时保留私密性的元素。"

在1974年设计马恩拉瓦莱的"绿色之塔"后，鲍赞巴克将视野转向巴黎的一组150单元住宅设计竞赛，基地中原有的拉·若凯特女子监狱被拆除。他的方案立即被视为一个宣言，因为它提供了联系建筑与城市的新途径。他将场地转换为原型式的自然地带——植物的纪念碑——嵌入城市的心脏。此处，虚空被作为形式。纯粹的矩形平面在转角处打开豁口，每座公寓都拥有可供选择的视觉景观和绝佳的自然采光。

1974年鲍赞巴克才过而立之年。此前他名下仅有一座单体建筑，但这次的新项目使他着手去明确未来的理论与实践中将要采取的形式语言。同时他也为城市规划师与社会学家做出了榜样，在满足既定密度的同时——150单元——完全突破现代主义的底限，将其置于彻头彻尾的从属身份，让位于空隙与"迷失"的空间。

拉·若凯特方案是理论与法则，而5年后的欧风路项目则是法则的应用。他关于城市建筑的一切思考与探究都曝露于此，这证明城市中的集合住宅能够具备出类拔萃的建筑品质，而公共空间也同样能够亲切宜人，而非注定成为苍白麻木、反应迟钝的真空地带。

拉·若凯特并不是连续、重复不变的住宅，每所公寓都独具特征，合于建筑的特定位置。在这个位于巴黎13区的项目中，鲍赞巴克的二元性展现无余：私密与深邃辽远的视野并行不悖。为此，他竭尽所能创造出精彩绝伦的花园印象，而非社会住宅集合体，并与此配合，系统性地发展了窗牖设计模式，为220所公寓提供了1000余扇窗户。

鲍氏享用着巴黎令人崇敬备至的建筑盛宴，并思忖自己的设计模式，终于灵光一现，大功告成。新闻媒体矫枉过正，推

欧风路社区项目塑造了两个（不同）时代间的和睦与亲善，在巴黎和世界建筑业界引起广泛关注。
A rapprochement of two periods, the Hautes Formes neighborhood project drew enormous attention in Paris and from professionals all over the world .

childhood is there in all his housing projects: in his way of mingling indoors and outdoors, making interior and exterior equally important and presenting dream and reality. Once again he's creating landscapes, vistas, distant perspectives, lines of flight.

"I like to bring out the various elements in a building," he explains, "and this leads to an interest in contrast and duality. The beauty of something supple, as I see it, is revealed by a rigidity to be found elsewhere. I like generating contrasts, complementarities, perceptual duels...I like working big and expansive, while retaining an element of intimacy."

After delivering his "Green Tower" in Marne-la-Vallée in 1974, Portzamparc set his sights on the competition for a group of 150 housing units on the area freed by the demolition of the La Roquette women's prison, in Paris. His proposal was immediately taken as a manifesto, in that it paved the way for a new city/architecture connection. He transformed the site into a primal nature zone — a vegetal monument — carved into the heart of the city. Here emptiness is recognized as form. Open at each corner, this pure parallelepiped provides each apartment with a choice of views and excellent natural lighting.

In 1974 Christian de Portzamparc had only just turned thirty. He had a single building to his name, but with this new project he set about making clear what form the coming theory and practice would take. He also turned it into a lesson for the civil service town planners and sociologists by meeting the required density level — 150 units — at the same time as he seriously undercut the current modernist line, which accorded a purely subordinate status to interstitial and "lost" spaces.

La Roquette was thus the theorem of which, five years later, the Hautes Formes project would be the application. All the elements of his thinking about urban building are present here as the proof that multiple housing in the city can have real architectural quality and that public space can be habitable, rather than simply an unresponsive vacuum.

La Roquette is not serial housing, for each apartment has characteristics corresponding to its specific position in the building. Here in the 13th arrondissement Portzamparc's dualism is clear: privacy combined with views into the distance. To achieve this, he pulls off the stunning feat of creating the impression of a garden rather than a social housing complex and of inventing a combinative system of

"相衔创作法"还是精致优雅、活活脱脱的实体？无论如何，福冈项目是拼贴法应用于城市布景透视的杰出实例。

"Cadavre exquis" or exquisite living body? Whatever, the Fukuoka ensemble is a remarkable example of collage applied to urban scenography.

崇其具有“革命性”，但这使鲍赞巴克颇感欣慰，并且追思起2500多年前先哲老子的箴言：“凿户牖以为室，当其无，有室之用。故有之以为利，无之以为用”。

10年之后，鲍赞巴克接受地区首席建筑师矶崎新的邀请到日本福冈一展身手。被邀者的名单令人振奋，美国的斯蒂芬·霍尔，荷兰的雷姆·库哈斯，澳裔加利福尼亚建筑师马克·麦克，和西班牙的奥斯卡·塔斯奎茨。这座海滨城镇迅速膨胀、负载着千篇一律、比比皆是的颇具理性气质的板式建筑和塔楼，矶崎新有自己的构想，意图在这里造就活力四射，丰富多彩的邻里社区，于令人沮丧、繁复冗余的场地中塑造出沁人心脾的景色。他所希冀的实际上是超现实主义者的“相衔创作法”（cadavre exquis，也称美尸游戏——译

openings offering over 1000 windows for 220 apartments.

With the entire Paris architecture scene queuing up to admire and then to ponder its own approaches, this young architect had suddenly made it. Media overkill boosted him as a "revolutionary", which made him smile and think back to Lao Tse, 2500 years earlier: "My house is not the wall, it is not the floor, it is not the roof: it is the empty space between them, because that is where I live."

Ten years later Portzamparc found himself asked to Fukuoka in Japan by Arata Isozaki, chief architect for the district. The impressive guest list also included America's Steven Holl, Holland's Rem Koolhaas, the Australo-Californian Mark Mack and Spain's Oscar Tusquets. Isozaki's idea for this small, rapidly expanding coastal town already well laden with rational, repetitive slabs and towers, was to create an energetically diverse neighborhood, a touch of scenery in an

鲍赞巴克所有的志趣在这里融会贯通：雕塑，绘画，素描，建筑。

Here all Portzamparc's interests combine: sculpture, painting, drawing, architecture.

者注）——一系列偶然写就的元素——而鲍赞巴克把握住了机遇。

"你不应认为我仅仅在此地经历到颓唐与破败"，安德烈·布列东在《纳底雅》中写道，"我谈及的话题将限制在那些不期而遇、出乎意料的事物中；那些从天而降的，特定的喜爱与厌恶之物给予我的一切。我在讨论时不因循任何预先设定的秩序，仅仅让应该显露的显露出来而已。"

一如我们所见，鲍赞巴克的超现实主义更加关注方式方法，而不是记录与书写。他使得理应显露之物显露出来，使用令人

otherwise uniformly depressing setting. What he wanted, in fact, was a version of the Surrealists' "cadavre exquis" — a chance sequence of written elements — and Portzamparc pounced on the opportunity.

"You mustn't expect a complete rundown on what I've experienced in this field" wrote André Breton in Nadja (André Breton, Paris, Gallimard, 1928). "I shall limit myself to talking about what sometimes happened to me in a completely unlooked-for fashion; about what, reaching me by invisible pathways, gives me the measure of the special favors and disfavors granted me. I shall speak of this in no predetermined order, letting what comes to the surface come to the

表皮，色彩，韵律的交响。
Interplay of surface, color and rhythm…

空间开敞且阳光充足的公寓拥有辽远的视野，且常常会在两个方向上开敞。
Open, light-filled apartments, often with two aspects and always with a view.

欢愉的素描、绘画和雕塑的混合体：白色的建筑，绿色的花园，墨色的砾岩，赭石的坦比哀多，斑驳的地面，还有许多个人性的参照物，包括欧风路，欧罗迪斯尼，楠泰尔，波尔多的月之港等方案。

他的设计成为精致优雅、活活脱脱的实体，而不仅仅是简洁细腻的"相衔创作"作品，它们蕴藏着另一巴洛克艺术策略：将城市作为舞台的布景透视法——师承自巴洛克教堂的经典手法。

90年代来临了，他主持了巴黎贝西区的一项包含67套公寓的住宅项目。基础图纸由建筑师让-皮埃尔·布非绘制——他同样是68年5月巴黎美术学院职业委员会的成员——他构思了规则的建筑体块，这些体块沿着新贝西区公园的边界排列成行，阳台一线延伸，稳重可靠，内庭院则朝向公园微微开敞，并将街道阻隔其外。

在新贝西地区闲庭漫步，人们会被布非方案的精巧与不折不扣的城市气质所感染。而且，方案成功融入到这个强调个体而非社会整体的时代。布非再次邀请了一支人才济济的建筑师团队——亨利·奇里亚尼，弗兰克，雷克勒和杜萨品，伊夫·里昂，费尔南多·蒙特斯和其他一些人——深化构思，勾勒具体形式。

鲍赞巴克惟一的难题是布非的规划无法让他尽兴地敞开建筑体块。然而机遇和必要之手段常伴其侧：一棵幸运的受到保护的栗树成为与规则洽商博弈的砝码，他得以雕琢建筑形体，使其与公园之间拥有最佳的视角，他还在北立面上设置一处豁口，将建筑一分为二。更广阔的视野，更明媚的阳光——鲍赞巴克娴熟地处理楼层空间的分隔与角度，卓有成效地获得更多自然光。

1998年的格勒诺布尔展现了全然不同的经济背景：一项低投入的住宅项目要求进行详尽的规划设计。鲍氏以那些名不见经传的当地建筑为原型，重组整个社区，其布局自成一体，塑造了多边形开放平面的街区：最重要的是，这里有总体性的组织，简洁的形态和山形的屋顶，而活灵活现的着色粉刷让人联翩浮想起四五十年代的连环漫画。

城市景色映衬着公园的美景：贝西区公寓贡献了视觉的饕餮大餐。

View of the city plus view of the park: the Bercy apartments offer the eye endless opportunities.

交叉口，走廊，连接处和人行桥，它们依然未显过时陈旧之态。

Crossings, passageways, connections and footbridges still very present.

surface."

As it turned out, Portzamparc's use of Surrealism had more to do with method than with writing. He let what came to the surface come to the surface in a joyous mix of drawing, painting and sculpture: white architecture, green garden, black rock, ocher tempietto, mottled flooring and personal references including Hautes Formes, EuroDisneyland, Nanterre and the Port de la Lune in Bordeaux.

Much more an exquisite living body than an exquisite cadaver, his project involved yet another Baroque strategy: that of urban scenography and the city as stage — the primary lesson, in fact, offered by the Baroque cathedral.

Then it was back to Paris as the 90s opened with a brief for 67 apartments in Bercy. The basic charter drawn up by architect Jean-Pierre Buffi — another ally from the Ecole des Beaux-Arts occupation committee in May 68 — called for an alignment of regular blocks along the edge of the new Bercy Park, with stable, linear balcony heights and internal courtyards opening slightly onto the park and shutting out the street behind. The Buffi plan was highly intelligent and thoroughly urban, as a stroll around the new Bercy makes clear. It fits, moreover, with a period that tends to stress the individual rather than society as a whole. Once again, Buffi called on an imposing team of architects — Henri Ciriani, Franck Hammoutène, Leclerc and Dusapin, Yves Lyon, Fernando Montès and others — to give his project form.

The only problem for Portzamparc was that the Buffi charter didn't let him open up his block as much as he would have liked. However, chance and necessity were on his side: a providential protected chestnut tree provided a way around the rules, letting him sculpt building-objects offering the best possible views of the park and open a breach in the north facade, thus cutting the building in two. Broader vistas, more light — Portzamparc made skillful play with angles and stories to bring the sky in more effectively.

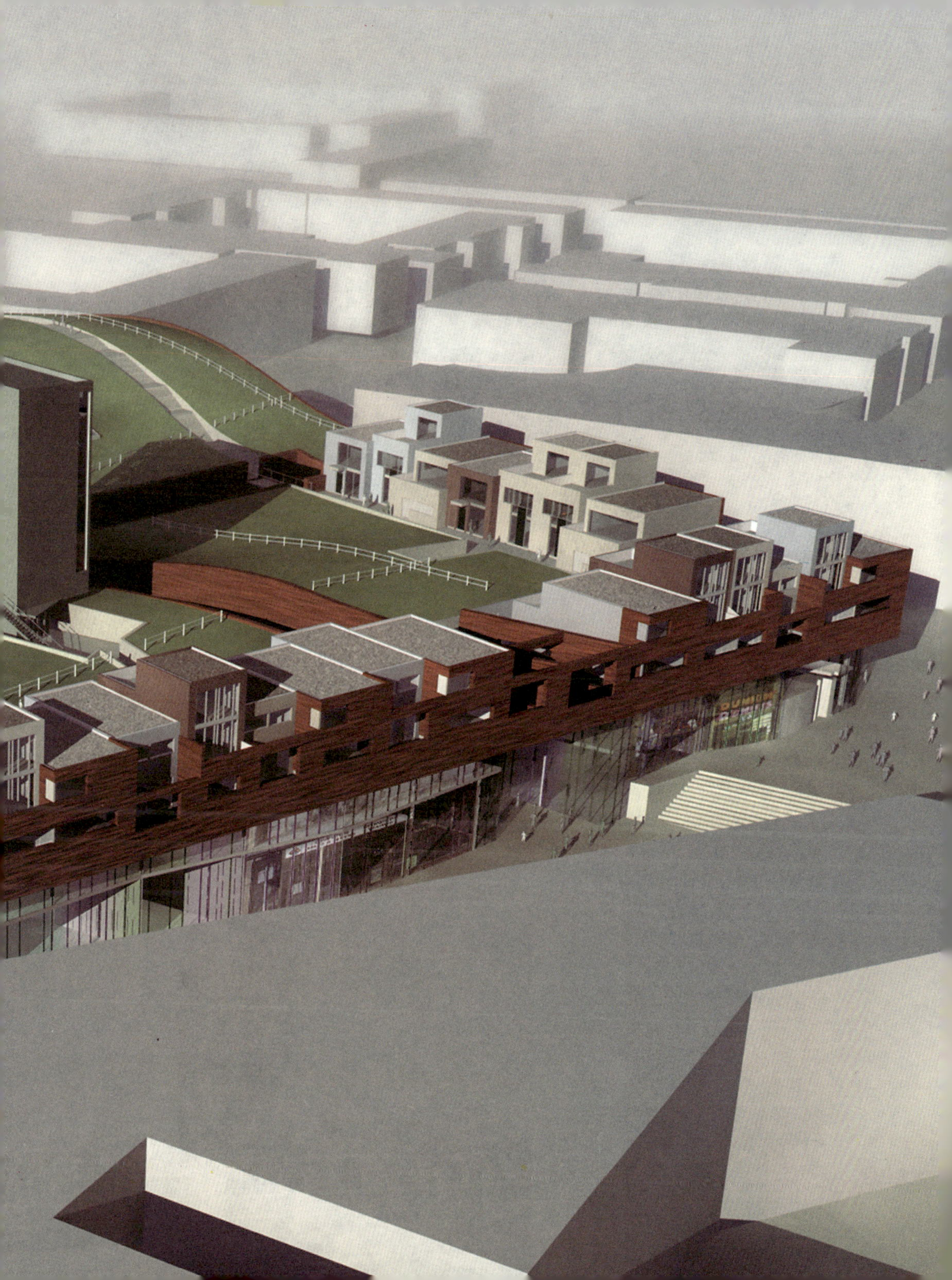

阿尔梅勒市，荷兰：雷姆·库哈斯希望市区采用"千层酥饼"的设计模式，鲍赞巴克作为被邀请的建筑师之一，手到擒来，随心所欲地诠释这一理念。
Almere, Holland: Rem Koolhaas wants a "flaky pastry" approach to the downtown area. As one of the invited architects, Portzamparc interprets the idea very freely. .

在荷兰的阿尔梅勒市，鲍赞巴克目前正从事着全新的"特技表演"。阿尔梅勒坐落于围海造田的低地上——填海开垦的土地——是一座欢腾雀跃的新城，其规划承袭传统理念，讲究"条条大路通罗马"的气韵。

在首席建筑师雷姆·库哈斯的统率下，要进行的任务是将城市建造完成——建设与常规模式格格不入的城市市区。作为达此共识的努力之一，鲍赞巴克的方案采纳了较为中心化的建筑体量，抬高的区域下部为道路和停车场库，上部则留给步行者，商铺和住宅所用。

紧握基地——他称其为"魔术地毯"——的禀赋与机遇，鲍赞巴克扭曲并粉碎了意料中的直角形态，创造出突兀效果，使各个交叉点位移变更：一切都是为

Grenoble in 1998 represented a different economic context: a low cost housing venture requiring a detailed urban plan. Beginning with a corner building prototype, Portzamparc reorganized a whole neighborhood according to a layout that generated its own masses and shaped polygonal open-plan blocks: all in all, generic organization, simple shapes, mountain-style roofs, and vivid renders that recall nothing so much as the comics of the 40s and 50s.

At Almere in Holland, he is currently engaging in fresh acrobatics. Set in the polders — land reclaimed from the sea — Almere is a dynamic new town planned on classical lines, where "all roads lead to Rome".

The task, under chief architect Rem Koolhaas, is to finish the city — and to finish it via the downtown area, in direct contrast with the way things have always been done. As part of this concerted effort, Portzamparc has inherited a very central block, a raised area that will have roads and parking lots underneath, with its upper part set aside for pedestrians, stores and housing.

Seizing the opportunity of a site he calls his "magic carpet", Portzamparc has twisted it, shattered its anticipated right-angledness, created projections and shifted the intersection: all in the interests of long-distance views. Yet again, from Hautes Formes to Almere and from Fukuoka to Bercy, the Portzamparc style seems to cut back and forth, opening out across all sorts of registers and blossoming in its use of the different.

The most convincing judgement of all, in the final analysis, comes from Ada Louise Huxtable, architectural historian and redoubtable critic for the New York Times: "Without exception the buildings designed by Christian de Portzamparc are the expression of an architecture full of joy, one that far outstrips the rigidity of modernism and the cardboard of postmodernism."

了获得绝佳的远距离视野。从欧风路到阿尔梅勒，从福冈到贝西区，鲍赞巴克的设计风格左右逢源，不拒斥任何记录在案的经典之作，在五花八门的变革中日臻成熟，大放异彩。

对鲍氏最具说服力的判词来自建筑历史学家和令人敬畏的批评家安达·路易斯·赫克斯泰勃尔，他在纽约时报上撰述道："鲍赞巴克的设计无一例外诠释了满怀愉悦的建筑学，它们远远超越了现代主义刻板僵死的繁文缛节和后现代主义脆弱肤浅的'金科玉律'。"

建筑师在这里探究开拓场地，关注透视问题，最终创造出举世无双的景观效果。

Here the architect exploits the sweep of the ground and plays on perspective to create a unique landscape.

聚集
coming together
7

公共空间，公共场所

Public spaces, common places

“咖啡馆内，姑娘们志趣盎然。夜晚，她们徘徊良久，然后健步登上一位建筑师专为她们设计的楼梯——忽然之间，她们的双腿变得轻松愉悦起来。”（Stéphane Guibourgé, Citronnade, Le Dilettante, Paris, 1991）

这场景当然发生在波布尔咖啡馆里。Stéphane Guibourgé 仅用寥寥数语就刻画出克里斯蒂安·德·鲍赞巴克的这一作品。更为精明难得的是，他觉察到鲍赞巴克的夫人凯瑟琳对鲍氏的影响，而且发现了这位建筑师在楼梯、涡卷、螺旋形体，以及陡坡形态设计上的杰出才能。鲍赞巴克的

"The girls in the cafes are amusing. In the evening they walk about a little, climb a flight of stairs lovingly designed for them by an architect — and all of a sudden their legs become a pure delight" (Stéphane Guibourgé, *Citronnade*, Le Dilettante, Paris, 1991).

The scene takes place, of course, at the Café Beaubourg, and in one brief paragraph Stéphane Guibourgé pinpoints Christian de Portzamparc's work there. More astutely still, he detects the influence of Portzamparc's wife Catherine and senses the architect's special talent for staircases, scrolls, spirals and ascents.

As it happens, a remark by Portzamparc echoes Guibourgé's observation: "Some evenings you can't find a seat. There's the lofty nave area,

眩目绝伦的波布尔咖啡馆中厅。高处的架空连接桥意味深长。

The stunning nave of the Café Beaubourg, where the overhead walkway plays a vital role.

夹层：清寂静谧，且/或可以俯瞰楼下。

The mezzanine: solitude and/or the chance to observe what's happening below.

一次谈话恰好印证了的观察："在一些夜晚，你甚至找不到座位。巍峨的中庭空间适合喜欢在社会生活中崭露头角的人；凹进的壁廊更适宜喜欢幽静氛围的一族人；夹层则供人们俯看市井风情，写些东西或者静静地读报：一切正如我所愿，人们找到了在咖啡馆内所需的东西。此外我还在听觉效果、照明、内部材料，以及家具设计等许多方面给予了极大关注"。

关键的问题是要创造一个形形色色的人群都愿意涉足的场所，吸引他们更深层次地进入并体验这座建筑。"记得参观圣米歇尔山时，我在台阶上上下穿梭，惊异于各式的动人发现和多种多样的流线选择。"他具有能够使用几句简单明了的词语概括游客的所想与所需的天赋：惊奇、变化、空间序列与视野，这些元素使你更远、更高、更深地进入建筑。这种种的过渡与变化几乎可被视为文学作品——"这就像讲故事，你须倾听情节的叙述，直到关键的'包袱'（即故事高潮或相声的逗笑处——译者注）被抖出"——但这需要高度的个人敏感性："建筑学不是简单的智力问题。不可能像我们自1968年以来

for people wanting to be visible on the social scene, the niches for the more retiring types, the mezzanine for looking down on the show, writing a letter or quietly reading the paper: it all worked very quickly the way I intended it to, with people finding just what they wanted in a cafe. And I'd put a lot of care into the acoustics, the lighting, the interior materials, designing the furniture and so on."

The big thing was to create a place all sorts of people would want to go into, to draw them further into the building. "I remember visiting Mont St Michel, climbing stairs, descending stairs and being knocked out by all the things to discover, all the possible itineraries." He has the gift of pulling everything together in a few apparently commonplace words that somehow sum up exactly what a visitor wants: the surprises, changes, sequences, views that take you further, higher, deeper. Once again the ongoing transitions are seen in an almost literary way — "It's like a narrative that has to keep you moving until you get to the punchline" — but one informed by a highly individual sensibility: "Architecture's not simply a matter of intellect. It can't be reduced to words, the way we thought it could in the years following 1968. It's like dance: just as dancers' movements create space on the stage, somebody's movements through

布戴尔美术馆的精巧与优雅：怎样使完全封闭的空间开敞起来。

Subtlety and elegance at the Musée Bourdelle: how to open up a totally closed space.

的这些年认为的那样，简化为只言片语。它仿佛是舞蹈：正像舞蹈家的动作在舞台上创造空间一样，人们在建筑中的运动与游历，揭示并彰显了建筑本身。”当你进入布戴尔博物馆，体验它简洁朴素而又灵巧优雅的平面交响时，立即就会对这一论述信以为真：白色石制地面捕获并反射倾斜的光线，使正午时分刺眼的日光变得柔和，能够吸收光照的灰色水泥墙面为青铜雕塑提供了背景。贯穿这一建筑形式的还有另一更为关键的首要形式主题——即一系列拥有妖娆光线的围合空间——人们被吸引、邀请、引领着，从一尊雕塑，一幅绘画走向下一件作品。

在设计布戴尔博物馆扩建部分的同时，鲍赞巴克还在为欧洲迪斯尼构思一座饭店。这被很多人认为——包括批评家菲莉斯·兰伯特——完全是对音乐城的回应，因为它们都是苦行禁欲气质与幻想虚空韵味的结合体：“你们正在做前无古人之事！”

鲍赞巴克的欧洲迪斯尼饭店，是表现力、色彩、与幻想的结合体——也是对里约热内卢海湾的某种隐喻。
Portzamparc's EuroDisneyland hotel project combines expressiveness, color and fantasy - while offering a kind of metaphor of the bay of Rio de Janeiro.

a building reveal and recount its architecture to him." You don't doubt this judgement for an instant once you're inside the Musée Bourdelle, with all the intense simplicity and subtle elegance of its planar interplay: a white stone floor to capture and reflect oblique light while attenuating the brightness of noon, and gray cement walls to absorb light and provide a backdrop to the bronzes. And moving through this secondary architectural form set into the first — a succession of enclosed spaces whose lighting is totally spellbinding — one feels drawn, invited, guided from one sculpture, one drawing to the next.

While preparing his addition to the Musée Bourdelle, Portzamparc was also working on a hotel for EuroDisneyland, a similar combination of ascesis and fantasy that elicited from many — including critic Phyllis Lambert — a reaction based solely on the Cité de la Musique: "You're doing something new here!" Granted, understanding him from one project to the next is no easy task: there's a solid, well-founded internal consistency, but it comes wearing a

毋庸置疑，一个接一个地理解他的所有设计方案并非易事：它们确实存在根深蒂固的内在一致性，但这却隐藏于外表多变的面具下。鲍赞巴克若无其事却活力充沛地解释道："我喜欢这项工作的真正原因是每个项目和场地都会成为全新的挑战。每一次都有探索发现的机会，而我自然而然地倾向于保持运动，着迷于探索……"

出其不意，他不再步索莱尔，德勒兹或巴特思想的后尘，转而师从弗朗西斯·皮卡比亚："你必须具备流浪气质，将思想当作城市和街道，徜徉穿行其间。"就欧洲迪斯尼饭店而言，这种横贯与穿行可非同小可。项目的第一次会晤在巴黎大饭店举行，与会者有迈克尔·埃斯纳及其下属员工，还包括全体建筑师：盖里，格雷夫斯，霍莱因，库哈斯，文丘里，努韦尔和格伦巴赫。氛围颇有些古灵精怪，但这有何不可呢？鲍赞巴克的方案是"美洲式"的——但毫无疑问属于南美风格：它是对里约热内卢海湾的隐喻，人们很难猜测这一隐喻体矗立在巴黎东部红色甜菜种植平原上的效果，但它使我们联想起中世纪的绘画和儿时摆弄的立体模型，它也预示着福冈的"巨石"方案和棱柱形态的纽约路易威登大厦。

与此同时，在壮丽而隆重的氛围中，另一次对幻想的探险在伊曼纽尔·温加罗服装专卖店的设计中得以进行，建筑坐落在巴黎的蒙田大道上，也是一系列设计项目的原型。温加罗是一位狂热的音乐爱好者，因为巴黎音乐城的缘故结识了鲍氏。而伊丽莎白恰巧又是温加罗时装风格的忠实追随者，于是，被一些人冠以"冷峻"、"保守"之名的建筑师鲍赞巴克决定放松身心，休整娱乐，然后精力充沛地驳斥那些怀疑他的人。设计成果是整整500平方米的可被感知并直接触及的空间，排列整

位于蒙田大道的温加罗服装专卖店是奢华与优雅世界的比例模型。

The Ungaro boutique on the Avenue Montaigne is a scale model of a world of sumptuous refinement.

椭圆形体的厚重肃穆传达了(某种) 信号，这里容纳了商业法庭。

Grasse: an impossibly steep site squeezed by winding roads. The solution was to make the difficulties a positive factor.

格拉斯：一处不可思议的陡峭场地，被周边的道路围绕积压。解决之道是将困难转化为积极性因素。
The massive solemnity of the oval section works as a signal and houses the commercial court.

mask, as Portzamparc explains with nonchalant brio: "What I really like about this job is the newness of each project and each site. Every time it's a chance to discover things, and since I'm naturally inclined to keep on the move, to keep exploring..."

Here, all of a sudden, he is no longer following in the footsteps of Sollers, Deleuze or Barthes, but of Francis Picabia: "You have to be nomadic, traversing ideas as if they were cities and streets."

In the case of the EuroDisneyland hotel, the traversal was no small matter. The first meeting happened at the Grand Hotel in Paris, with Michael Eisner, his staff and all the architects: Gehry, Graves, Hollein, Koolhaas, Venturi, Nouvel and Grumbach. An odd ambience, but why not? Portzamparc's project emerged as "American" — but, incorrigibly, South American: a metaphor of the Bay of Rio. A metaphor it was hard to imagine planted in the red beet plain east of Paris, but which suggests medieval painting and the dioramas we played with as kids, at the same time as it foreshadowed the Fukuoka "rock" and the prismatic geometries of the LVMH tower in New York.

Simultaneously, amid great pomp and ceremony, another venture into fantasy was realized in the form of an Emmanuel Ungaro boutique, the prototype of a long series, on Avenue Montaigne in Paris. Ungaro is a passionate music lover and was led to Portzamparc by the Cité de la Musique. As it happened, Elizabeth is a convinced "Ungarian", and the architect some describe as "cold" and "uptight", decided to cut loose, have some fun and vigorously refute the doubters. The result was 500 square meters of sensuality and tactility, with rows of boudoirs and dressing rooms on a dazzling floor of white Yugoslavian granite, the most precious of its kind in the world. Basically, of course, the project was a tribute to Elizabeth, his muse, whose body he transformed into a sculpture and made the focal point of the entire space.

The competition for the Law Courts in Bordeaux — ultimately lost to Richard Rogers — put Portzamparc on a steep learning curve in terms

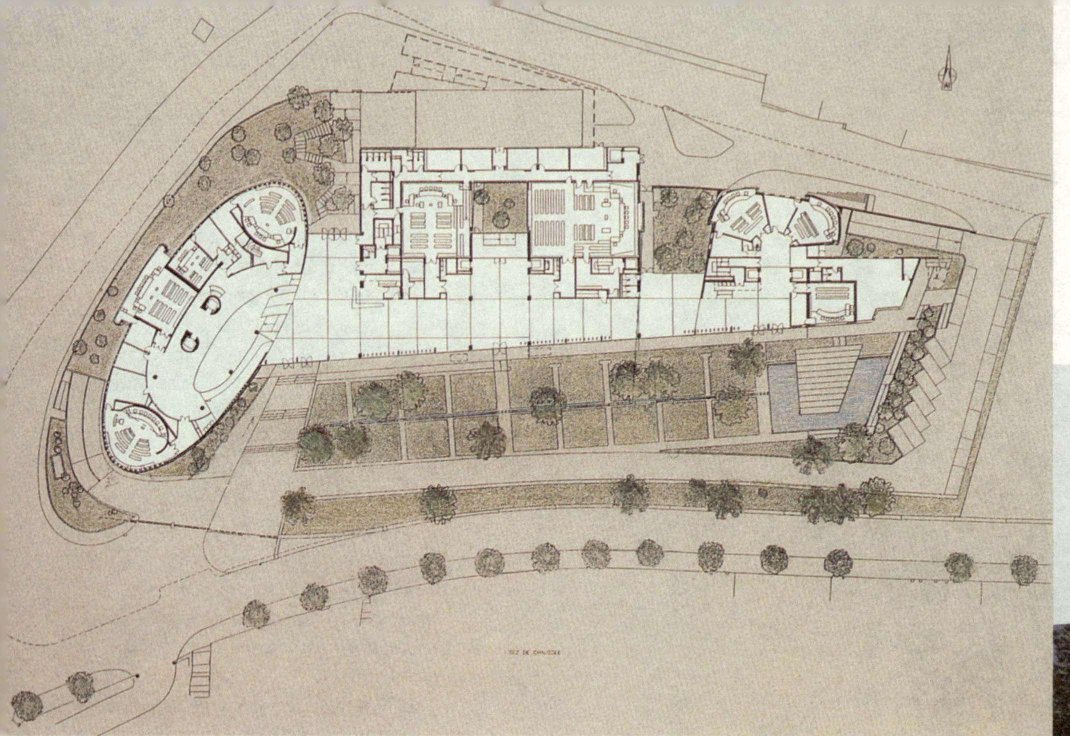

齐的化妆间和更衣室用世界上最为名贵的南斯拉夫白色花岗石铺装地面。当然，这项设计也是鲍氏送给自己缪斯女神——伊丽莎白——的礼物，他将妻子的身体变形为一尊雕塑，作为全部空间的焦点（系指服装店中心的造型雕塑——译者注）。

波尔多法院竞赛——最终输给了理查德·罗杰斯——迫使鲍赞巴克埋头制定研究司法工作模式的学习进度表：包括礼仪礼节、利害关系、其他各类需求，以及如何划分区域等等。有这些经验作坚实后盾，他参加了 1993 年的格拉斯法院设计竞赛。坦率地说，基地造成的困难让人进退维谷：场地坐落在倾斜的坡地上，被周边的道路围绕积压。就连那些声称每处新基地都令人欣喜的人，也会在此面露难色；再有，格拉斯是法国坡度最大的城市——这也是它诱人的一面——而且不论你身居何处，都能瞥见法院那意大利风格的屋顶。

of the way Justice works, its ceremonial side, its stakes, needs, and compartmentalizations. With this experience behind him, then, he took out the competition for the Law Courts in Grasse in 1993. The site was frankly impossible: on a slope squeezed about by winding roads. Even for the guy who says every new site is a delight, this one was a problem; but then again, Grasse is the steepest city in France — that's part of its charm — and wherever you are you're going to see the Italianate roof of the Courthouse. After some thought, the solution turned out to be simple and, as ever, inherent in the site and its function: Portzamparc decided it would be just as easy to enter from above as from below, with the visitor finding himself at once in the interspace of hall. This is the section that "aspirates" him, draws him in, then lets him proceed to a second zone of three readily intelligible geometrical approaches: solemnity for the oval building housing the commercial court, peace for the civil court and strictness for the criminal court. The hall is all the more "aspirating" in that it is perfectly transparent

整座建筑由三部分组成：经济法庭、民事法庭、刑事法庭。完全通透的大厅是统帅性要素。

The whole is made of three parts: the commercial, civil and criminal courts. The totally transparent lobby is the unifying factor.

空间的标点符号：
柱体、竖杆、标枪。
Punctuation: pillars, shafts and assegais .

格拉斯为鲍赞巴克提供机遇，他可以实践自己喜爱的带纹理的混凝土。
Grasse gave Portzamparc the chance to try out the textured concrete he loves.

一座审判庭的室内由伊丽莎白设计。
A courtroom fitted out by Elizabeth de Portzamparc.

一番思考之后，问题趋于简化，即一如既往地继承基地的内在禀赋与功能安排。鲍赞巴克决定建筑在不同标高处应该具有同样良好的可进入性，来访者会在不经意间已然身处大厅空间之中。这部分空间"吸入"和汲取人流，并引导人们继续向前进入三个不同几何形态、易于辨识的次级空间：椭圆形的经济法庭庄严肃穆，而民事法庭安宁和睦，刑事法庭则有几分狞厉之感。这座建筑的大厅通体透明，拥有最大化的"吸入"功能——这一特色远可追溯到楠泰尔歌剧舞蹈学校。正是在格拉斯，为了与土地和普罗旺斯的天空和谐共生，鲍赞巴克精心设计了带有纹理的混凝土——后来曾在柏林再次应用。同一年，他还完成了巴黎马约门的会议中心扩建项目。

"这里是一个与人会晤的滑稽有趣的地方"。马约门是巴黎众多门户中最不受欢迎的地方之一。这里荒凉萧条，令人沮

— a feature that links back to the Paris Opera Dance School at Nanterre. It was here in Grasse, seeking harmony with the earth and sky of Provence, that Portzamparc tried out the textured concrete that would appear again in Berlin. And in the same year he delivered the extension to the Porte Maillot Convention Center in Paris.

"Funny place for meeting people." Of all Paris's gates, Porte Maillot is one of the least welcoming. The area is desolate and depressing. The Convention Center itself is totally characterless — a building without architecture — yet it is endlessly in demand, whence the need for an extension. His thinking about matters urban led Portzamparc to work on the forecourt, density, and the vertical splaying that would create more space for parking, shops, exhibition rooms and offices, while also allowing for a balcony offering a majestic view of the city. Another advantage of the splaying was that it solved the fire escape problem. As with the Conservatory at Porte de Pantin, Portzamparc gives here the impression of building a fortress, a rampart that the

DES CONGRES

室内与室外：会议中心的锥形结构建构并诠释了这一主题。

Indoors/outdoors: the cone of the Convention Center structures and describes.

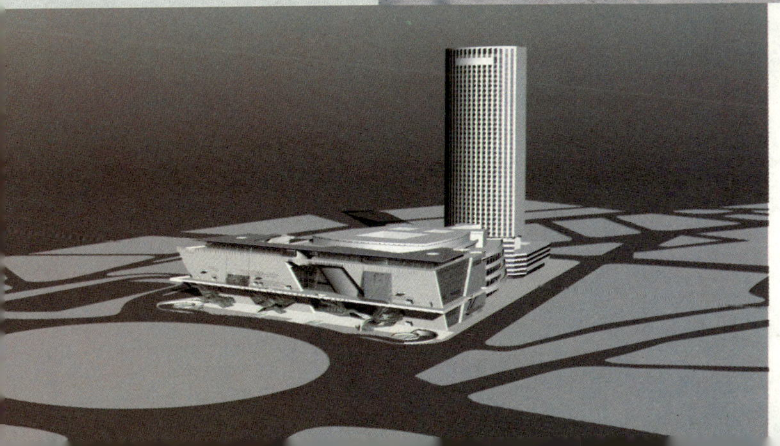

丧。会议中心本身毫无特色可言——一座没有建筑学的房子——然而却不可或缺，这也是扩建它的根源所在。鲍赞巴克关于城市问题的思考激发他对前庭与密度的构思，而竖直方向的扩张形态则创造出更多的停车空间，商店，展览厅和办公室，也成功塑造出一座用以观览秀丽城市风景的露台。竖向扩张形态的另一妙处是解决了火灾逃生问题。与位于潘廷门的音乐学校类似，鲍赞巴克试图在这里建造一座堡垒和城墙式的建筑，这引起建筑学期刊 Le Moniteur 的很具共鸣性却失之于偏颇与误导性的赞赏与膜拜，称其为“大手笔的立面”：在这立面之内，铺陈开一幅异常复杂缭乱的城市花格图样。建筑师成果的最佳概括性元素就是那座穿梭于室内与室外的锥体，它位于货物通过城门时征税加赋之地，猎户们昂首夸耀在布隆森林射杀的野兔战利品之时。锥体深深植入停车区域和地下铁路线，它在前庭中仿佛一扇旋转门，将来访者一扫而空，各自通达目的地。更为难得的是，它的上部容纳了一座500人音乐厅，轻巧而不对称的卷曲形态预演了位于基尔勃格的室内乐堂设计。这是当时此类艺术志趣的孤例，鲍赞巴克借此抒发了对阿尔瓦·阿尔托赫尔辛基芬兰大厦

餐厅，商铺，4000平方米办公空间，露台，500座音乐厅，停车场：会议中心是一座精密复杂的综合体。

Restaurants, shops, 4000 square meters of offices, terraces, a 500-seat concert hall, parking lots: the Convention Center was a complex, delicate operation.

architecture journal Le *Moniteur* resoundingly but misleadingly saluted as a "virtuoso facade": behind the facade, there lies an urban tracery of enormous complexity. But the element that best sums up the architect's work here is the cone, at once indoors and outdoors and set exactly on the spot where, when taxes were imposed on goods coming in through the gates of the city, hunters displayed rabbits shot in the Bois de Boulogne. Plunging down into the parking areas and even the Metro, the cone functions on the forecourt as a revolving door that sweeps up the visitor and bears him off. Better still, its upper section houses a 500-seat concert hall, lightly and asymmetrically scrolled in on itself in a foreshadowing of the chamber music room at Kirchberg. This was the sole artistic concern in this brief and Portzamparc presented it as a kind of homage to the Alvar Aalto of the Finlandia Hall in Helsinki.

The nocturnal illuminations for the facade of the Convention Center are a toned-down reminder of the experiments at the Bandai tower in Tokyo and the Pathé Cinema in Amsterdam. And maybe, too, the night view of the Strasbourg multiplex, where the luminous strip that enlivens the cornice generates an interplay of light and shade that looks like hanging fabric in movement; a touch which, incidentally, led to Portzamparc's first use of computers in an urban context.

1993 was a sumptuous year for Christian de Portzamparc: not only Grasse, but also the competition for a new cultural facility in Rennes that would unite the museum, library and science center, hitherto on three different sites. This was a moving experience: "I'd lived in Rennes in 1958-61 — an important time for me, because that was when I began to paint, started thinking about being an architect and finished

音乐厅位于锥体之巅，
微微蜷曲，且不对称。

*Set in the top of the cone,
the concert hall is
slightly scrolled and
asymmetrical.*

的敬仰之情。

会议中心立面的夜间照明成为他在日本东京万代大厦和荷兰阿姆斯特丹百代电影院设计实验的较为柔和的版本。或许与斯特拉斯堡复合体的夜景也有几分神似：闪耀的发光条带照亮了建筑檐口，它是光与影的协奏曲，仿佛悬浮空中动感十足的丝织品一般；这不期而遇的触动，引领鲍赞巴克第一次在城市背景中应用计算机技术。

1993 年是鲍赞巴克功成名就的一年：不仅在格拉斯，而且他还在雷恩参与了一座新文化馆的设计竞赛，这座建筑融博物馆，图书馆和科技中心为一体，这些功能在那时仍分散于不同的场地中。这是一次感人至深的体验："我曾于 1958 年到 1961

鲍赞巴克的会议中心是展现露台、窗牖、视野的画龙点睛之作。
Christian de Portzamparc's Convention Centre features a plethora of terraces, windows and views.

high school."

Another high point was the letter Frank Gehry, also a competitor, wrote to the client recommending Portzamparc's project. Whatever influence this may have had, the competition was won; but without knowing it, the architect had signed on for a long wait. Other urgent issues — the metro, for instance — were occupying Rennes' planners, but an unconcerned Portzamparc put the extra time to good use, rethinking and modifying. Out of the shilly-shallying emerged a conviction: the three elements making up the building had to be clearly differentiated, immediately intelligible and perceptible from outside. In no case were they to fuse into a single big box with stories for the library, stories for the museum and stories for the science center. "The museum had to be horizontal, so it became a plate lifting off

支撑在架空柱上的巨大形体，自地面漂浮起来，被另外两个体块打断：铝合金的圆锥体，和金属与玻璃构成的金字塔形棱柱体。雷恩的方案同时具备艰深的复杂性与彻底的易懂性。

An enormous volume on pilotis, apparently levitating over the esplanade, is punctuated by two others: an aluminum cone and a pyramidal metal and glass prism. The Rennes project combines great complexity with total intelligibility .

年期间居住在雷恩——这对于我来说是一个重要时期，当时我正开始着手绘画，考虑投身建筑师行当，并且完成了高中学业。”

另外值得一提的是，同为参赛者的弗兰克·盖里写信给业主，极力推荐鲍赞巴克的设计方案。无论这封信是否产生了影响，鲍氏最终赢得了竞赛；但出乎建筑师意料的是，他签约受雇但却仍要等待相当长的时间。其他的棘手问题——例如地铁——让雷恩的规划师们应接不暇，鲍赞巴克身为局外之人，充分利用这富裕的时间埋头思考与修改，从优柔寡断与犹豫不决中摆脱出来，作出决策：建筑的三个组成部分必须迥然不同，这应从外部便可感知和触及，它们决不能融合为一个巨大统一的形体，其中几层用作图书馆，几层容纳博物馆，另外几层分配给科技中心。“博物馆必须在水平方向展开，它支撑在底层架空柱上，仿佛从建筑首层腾空而起的雪玉冰盘，形态纯粹而巨硕，好似漂浮在地面之上一般。另外两座建筑体块插入这个封闭的混凝土形体：圆锥形的铝合金体块容纳科技中心，其上部球形的天文馆支配着整个体形；金属和玻璃构成的金字塔形棱柱体展开身形，直指天穹，内部作图书馆之用。通透迷人的首层空间仿佛供人穿行的城市邻里，设置三个公共出入口。我们看到三座截然不同的体块相互贯通，互相渗透，如孩童手中的玩具一般安装成形。”

与鲍赞巴克设计的其他公共区域一样，你可以在雷恩的方案中找到“峡谷”空间，即那些引导人们进入建筑内部的通廊，它们弯弯曲曲，分道扬镳之后复又汇合一处，希图建构无穷无尽的流线选择自由，并借此产生无穷无尽的情感。正像弗兰西斯·培根所说：“最炽热的激情是那些未经理智思考而直接进入大脑的情感。”

这座位于城市中心的独立建筑明确地区分出容纳其中的博物馆、图书馆和科技中心。

In the very heart of the city, a single building clearly states the distinction between the museum, the library and the science center it houses.

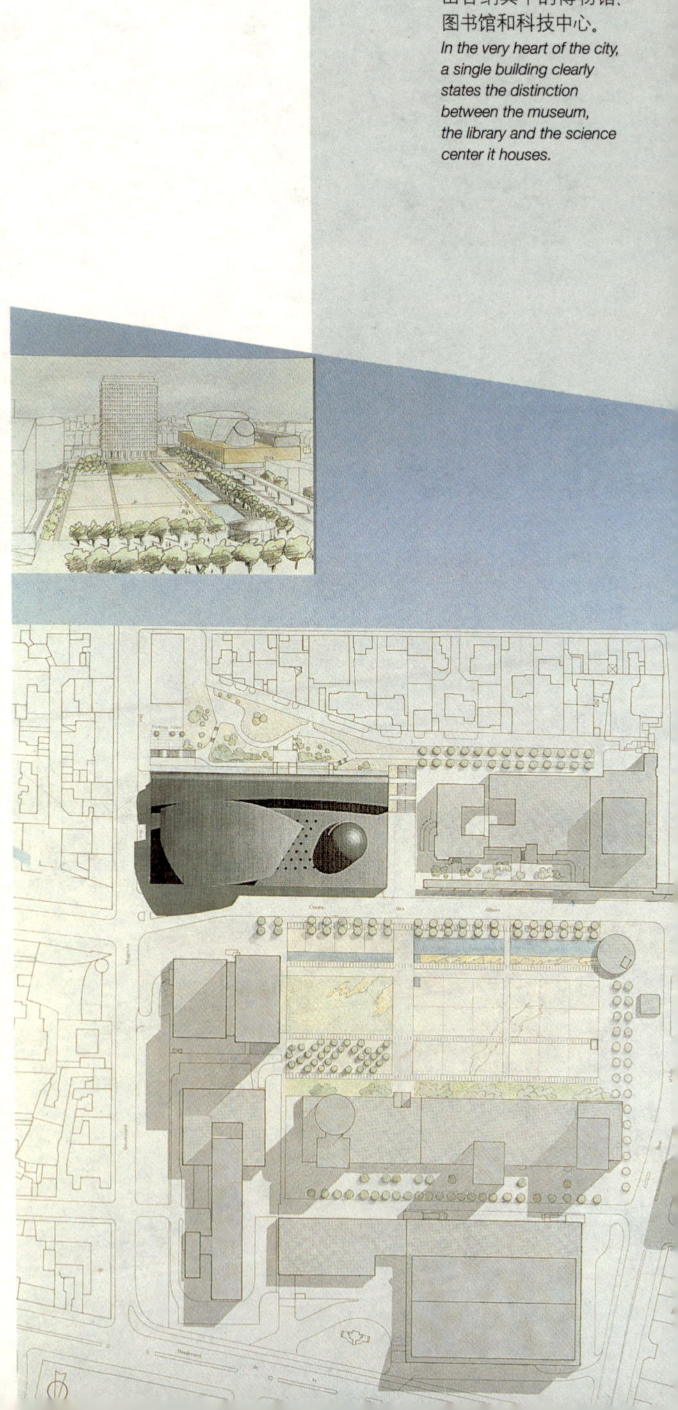

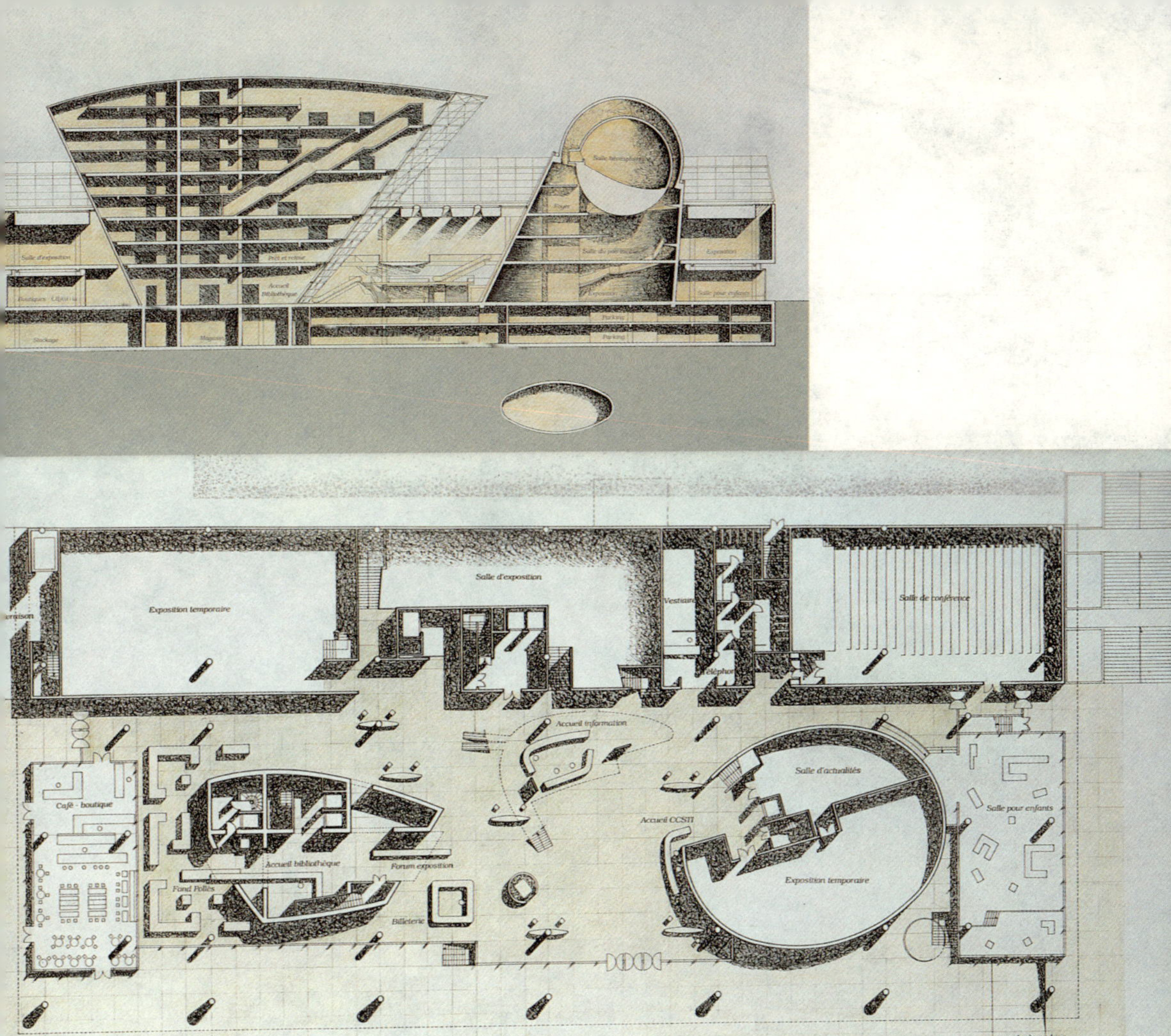

from the first floor, a pure, enormous volume on pilotis that seemed to be levitating over the esplanade. Two buildings cut through this concrete enclosure: the science center, with its conical aluminum volume dominated by the spherical planetarium; and the library, a pyramidal prism of metal and glass that opens out as it mounts skywards. Transparent and engaging, the first floor can be traversed like an urban neighborhood, with public entry at three different points. So you have three distinct shapes that intersect and interpenetrate, fitting together like the pieces of a kid's game."

In the Rennes project, as in all Portzamparc-designed public areas, you find "canyons", passages leading inwards that curve, branch off and come together again, reflecting that yearning for an infinity of itineraries generating an infinity of emotions. For as Francis Bacon said, "The most intense emotion is the one that reaches the brain without passing via the intellect."

立面上使用的红色混凝土成为表皮修饰、倾斜线条、与浮雕效果的突出特征，与马丁·华莱士共同设计。

The red concrete used for the façade features surface embellishments, obliques and reliefs designed in collaboration with Martin Wallace.

润饰
retouching

加伦河畔的月亮港，鲍赞巴克的构思是让城市设施在4千米的铁轨上自由移动，仿佛是无始无终的庆典一般。

The Port de la Lune, along the banks of the Garonne: Portzamparc envisaged urban equipment moving on 4 kilometers of rails in a kind of festival with no beginning and no end.

失之交臂
Missed opportunities

1980年代前期，伴随密特朗总统“大计划”的坚决推行，为数众多的设计竞赛此起彼伏。经过10年暴风骤雨般的营建和接下来20多年的资产大跃进，世界重新发现了建筑学，再次沉醉于表现力、意义和象征性的主题。更有甚者，法国成为样板，“法国学派”（French School）——折中主义与模糊性执拗任性的混合，热忱与创造力机智精明的共生——成为标榜对象被四处援引。

于是，设计竞赛成倍增加。但是竞赛的制度化往往使其沦为造星运动、拉拢选票、或天花乱坠的媒体宣传的纯粹借口。近30位鼎鼎有名的设计师常在被邀请之列，活跃在国际建筑舞台上。他们之中有来自英国的诺曼·福斯特和理查德·罗杰斯，来自荷兰的雷姆·库哈斯，来自意大利的马希米亚诺·福克萨斯和伦佐·皮亚诺，来自日本的安藤忠雄，来自美国的弗兰克·盖里和理查德·迈耶，还有来自法国的让·努

The early 1980s brought a profusion of competitions, with President Mitterrand's "big projects" moving forward hard and fast. After a decade of urgent rebuilding followed by two more of rampant property development, the world was rediscovering architecture, expressiveness, meaning and symbolism. And better still, France was becoming a model, with the "French School" — a heady mix of eclecticism and ambiguity, verve and inventiveness — being cited as an example everywhere.

So the competitions multiplied, but with a resultant institutionalization that too often made them mere excuses for star-making, electioneering and media hype. Thirty big names from the international scene kept cropping up on the list of invitees, among them Norman Foster and Richard Rogers from the UK, Rem Koolhaas from Holland, Massimiliano Fuksas and Renzo Piano from Italy, Tadao Ando from Japan, Frank Gehry and Richard Meier from the United States, and Jean Nouvel, Dominique Perrault and Christian de Portzamparc from France.

All of them were brimful of talent, with an

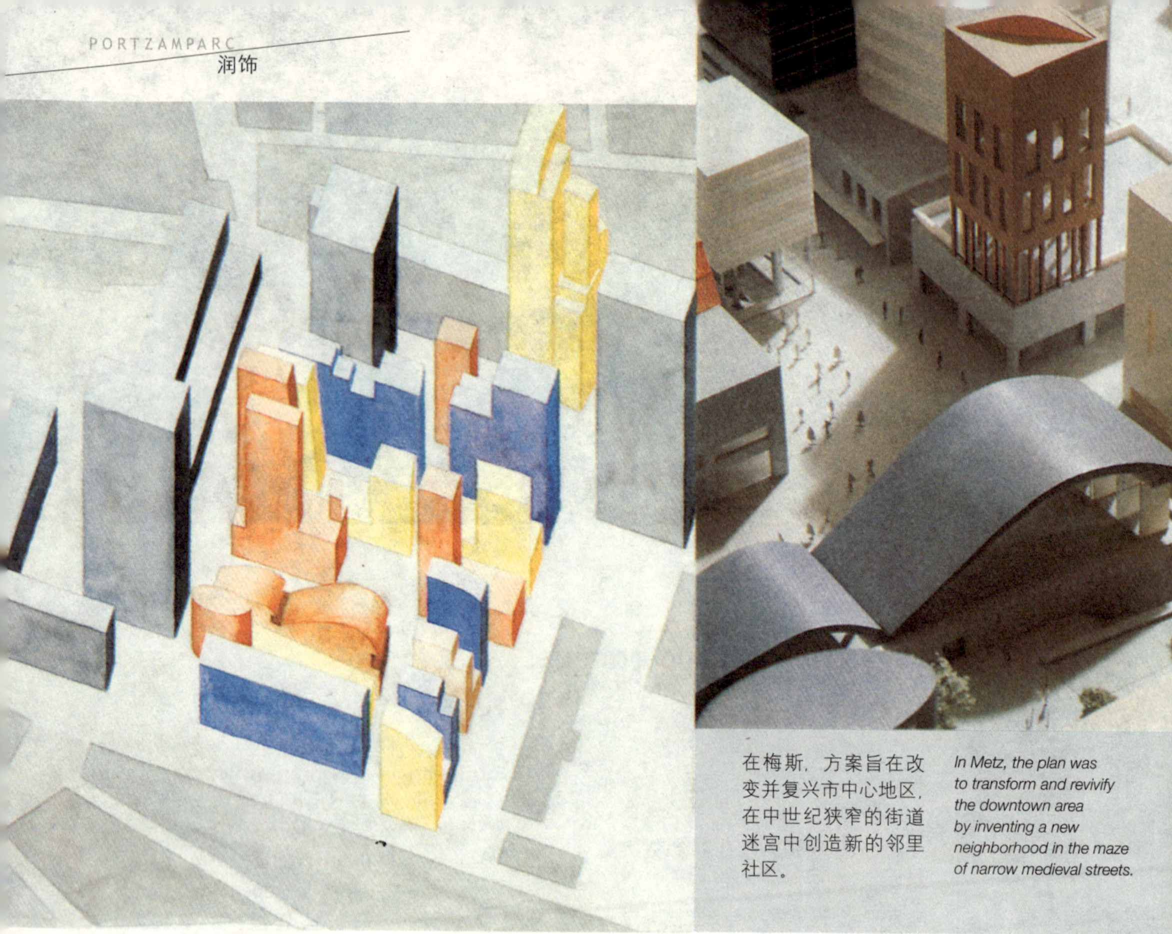

在梅斯，方案旨在改变并复兴市中心地区，在中世纪狭窄的街道迷宫中创造新的邻里社区。

In Metz, the plan was to transform and revivify the downtown area by inventing a new neighborhood in the maze of narrow medieval streets.

韦尔，多米尼克·佩罗和鲍赞巴克。

所有这些人都才华横溢，拥有人所敬仰的建筑学技巧与知识储备，创作风格琳琅满目，而且不乏超凡脱俗的感召力。但是，参与竞赛虽然意味着拥有胜出的机会，然而失败却总是十之八九。这失却了的形形色色的机遇常常给设计事务所带来经济上的危机。

克里斯蒂安·德·鲍赞巴克同样无法抵挡这滚滚洪流。他的档案袋里尘封着名目繁多的设计平面、方案模型和照片，出于种种坏的（经常）或好的（不经常）原因，这些方案从未得见天日。然而，这些资料与其说是遗憾与懊悔的记录，倒不如说是

arsenal including as much architectural knowhow you could wish for, the entire stylistic gamut and no lack of the necessary charisma. But the problem with competitions is that sometimes you win and often you lose, the result being all sorts of missed opportunities that can mean financial ruin for an agency.

Christian de Portzamparc was not immune to the syndrome, and his files contain a host of plans, models and pictures from projects that, for all sorts of (often) bad and (less often) good reasons, never saw the light of day. Yet these remain a source of retouchings rather than of regret or remorse: a creative idea does not necessarily have to be made concrete to exist, and all this research, imagination, experimentation and invention would enrich the

在拉维莱特湖与乌尔克运河的交汇处：梦寐以求的巴黎艺术学校。
Where the La Villette lake and the Canal de l'Ourcq meet: the art school Paris was dreaming about.

（对建筑师成就的）润饰：创造性的思想未必要依托业已实现的事物而存在，各式的研究、纷繁的想像、种种的实验与发明都将充实未来的工作。这些失之交臂的机遇包括一个“城市庆典”项目，两个开放式街区，两座学校，和三座文化类建筑。

巴枯宁曾梦想一次“无始又无终的庆典”。而这里提及的项目是波尔多月亮港4公里长的滨水城市地带。由Arc en Rêve主持的内部竞赛邀请库哈斯，努韦尔和鲍赞巴克等人参与；但结果却不了了之。鲍赞巴克的构思慧眼独具，利用现存4千米长的铁轨，作为展示／隐蔽庞大城市实体构筑物的手段，他希望借此保留城市的港口氛围，而这些实体将取代桅杆、林立的烟囱和隆隆的起重机；况且，这种设计模式还能在不来梅，鹿特丹，热那亚，那不勒斯，伊斯坦布尔或其他地区推广应用。这便是广袤无尽的“建筑庆典”，遵循被音乐家称为“无穷动”（moto perpetuo）乐曲的规律：波尔多的18世纪背景将亘古不变，而各类城市实体构筑物则周而复始地循环。

work to come. Among the missed opportunities were an urban celebration, two open blocks, two schools and three cultural buildings.

Bakunin once dreamed of a "celebration with no beginning and no end". The one in question here was to be four kilometers long on the wharves of the Port de la Lune in Bordeaux. In a restricted competition organized by Arc en Rêve, contenders included Koolhaas, Nouvel and Portzamparc; but nothing came of it. Portzamparc's approach was to use the 4 kilometers of existing rail track as an impressive way of revealing/concealing gigantic urban objects that would replace the masts, chimney stacks and cranes and so maintain the city's port atmosphere; furthermore, the operation was seen as reproducible in Bremen, Rotterdam, Genoa, Naples, Istanbul and elsewhere. This was to be an immense architectural celebration along the lines of what musicians call moto perpetuo: Bordeaux's 18th-century backdrop would live on changelessly, while the urban objects continued their round.

Metz 1990 and Frankfurt 1999: two open blocks. With the Place Coislin project in Metz, the task was to transform and enliven the downtown area, with its maze of medieval lanes. This was

1990 年的梅斯竞赛和 1999 年的法兰克福竞赛：两座开放式街区。梅斯的考斯林广场项目宗旨是在中世纪弄堂街巷的迷宫中对市区进行改造以激发其活力。这是特技一般的实践与演练，与法兰克福项目的尺度大相径庭，那里的项目是地区整合计划，远远超出设计任务书的要求。但梅斯项目最终石沉大海，而鲍氏又在法兰克福竞赛中败走麦城，两个方案最终成为永远不会被奏响的音符。

一个奇思妙想 从克劳德－尼古拉·勒杜设计的圆厅出发——这座建筑俯瞰拉维莱特湖，并且与艾里克·萨蒂音乐学校隐隐有些神似之处——你可以跟随摄影师罗伯特·杜瓦诺和作家安托万·布隆丹沿着乌尔克运河顺流而下。"运河指引你去往各处。对我来说，这里仿佛——而非他处——是属于我自己的世界，这感觉就像是沿着存在的边缘俯身前行……"（In Journ'haln 1 December 1984）

行走良久，在湖泊的尽头，灰暗阴霾的总督仓库伫立在那里，那是始建于 1850-1853 年间的建筑。它们在后来的年月中几经修缮，并随着牲畜交易市场和屠宰场的关闭腾搁一空，此后就一直沉寂闲置——直到成群结队的画家、舞台布景师、

空间韵律，走道，倩影：两座横跨运河的岛屿建筑。

Spatial scansion, walkways, reflections: two island-buildings bestride the canal.

an exercise in acrobatics on a totally different scale from the Frankfurt proposal, a district unification operation that went far beyond the brief. But with the project cancelled in Metz and the competition lost in Frankfurt, these were two scores that would never be performed.

A nice idea: starting out from the Claude-Nicolas Ledoux rotunda that overlooks the La Villette lake and has a sort of family resemblance to the Eric Satie Conservatory, you could follow the Ourcq Canal with photographer Robert Doisneau and writer Antoine Blondin: "The canal leads you everywhere: it seems to me now, that here and nowhere else should be my place in the world, with a feeling akin to that of following the edge of existence..." (In *Journ'hal* n° 1, December 1984).

Further along, at the end of the lake, stood the dark shapes of the Magasins Généraux, warehouses dating from 1850-1853. Altered and rebuilt over the years, then emptied with the closing of the cattle market and the abattoirs, they remained dormant — until a cohort of painters, scenographers, couturiers, photographers, interior decorators, musicians, graphic artists and architects moved in and brought them back to life. In no time at all 18,000 square meters of warehouse became 70 creative spaces; but then came the fire and one of the buildings went up in smoke, with its mass of wooden framing, beams and floors.

What was to be done? Put up a copy? Burn the

服装设计师、摄影师、室内装修设计师、音乐家、平面设计师和建筑师源源搬入，才给它们重新注入了活力。顷刻之间，整个仓库18000平方米的使用面积分隔成70个单元的创作空间；但后来其中一座拥有木结构框架、木制梁栋和实木地板的建筑不幸失了火。

该当如何呢？原样复制？还是将另外一座一并焚毁？构思应运而生：难道巴黎不需要一座本地的较大规模艺术学院吗？碰巧艺术评论家兼历史学家蒂埃里·德·迪弗设想了这个教育战略计划，而当时负责巴黎文化事务的让-雅克.阿拉贡也表示鼎力支持。但计划最终未能成行：克里斯

位于马恩拉瓦莱的建筑学校方案，旨在彰显本专业弥笃珍贵的多样性。

The school of architecture project at Marne-la-Vallée focused on giving shape to the sheer diversity of the profession.

other one down? The came the idea: didn't Paris need an art school, local but large-scale? As it happened art critic and historian Thierry de Duve had an educational strategy that Jean-Jacques Aillagon, in charge of Cultural Affairs for the City of Paris, wanted to see go through. It was not to be: Christian de Portzamparc entered the competition with a proposal that fitted him — and the City of Paris — like a glove; but sometimes, when the residents are up in arms, there's nothing anyone can do.

What he suggested was two "island-buildings"

蒂安·德·鲍赞巴克参赛并带来了投其所好的——也与巴黎这座城市匹配得天衣无缝——如同手套一般合体的方案；但居民们满怀敌意，对计划群起而攻之，任何人都只能白费心机。

鲍氏的构思是用两座“岛屿式建筑”替换原作，它们分立于运河两岸，被房间和跨越河道的平台走道相连接，与古斯塔夫·艾菲尔的金属拱桥遥相呼应。

蒂埃里·德·迪弗的设想触目惊心，结果自然也就更加令人失望了：课程学习分为两阶段，每个阶段耗时三年；第一阶段，对相关知识进行全面系统的学习；第二阶段则围绕独立的课题进行，学院24小时开放，成为探索、交流和实验的维系之地。蒙田在谈到此类教育模式时写道：“对学生的教育并非是去填满一尊花瓶，而是要去点燃燎原之火。”

鲍赞巴克在马恩拉瓦莱的建筑学院设计中——伯纳德·屈米在这次尚存异议的竞赛中折桂——再次采纳开放式学院的构思：三座独立建筑被杂树林和湖泊分隔开来，相互间靠通廊联系；它们坐落于共用的基座上，但视觉造型迥然不同。第一座具有集中式空间，以混凝土材料为特色；第二座采用离心式空间设计，金属材料特

在堪萨斯城的纳尔逊·阿特金斯博物馆扩建项目中，新建筑须与现状建筑相视而立。

For the extension to the Nelson Atkins Museum in Kansas City, the new building had to face the existing one.

to replace the originals, one on each side of the canal and connected by rooms and walkways crossing the water in a pleasing dialogue with Gustave Eiffel's metal swing bridge.

This outcome was even more disappointing in that Thierry de Duve's idea was terrific: two 3-year study programs, the first covering the whole learning range and the second focused on a project, in a school open as a twenty-four-hours-a-day junction for discovery, exchange and experimentation. Montaigne was talking about this kind of thing when he wrote, "Teaching a child is not filling a vase, it is lighting a fire."

Portzamparc's idea for the architecture school at Marne-la-Vallée — where Bernard Tschumi emerged victorious from what was yet another problematic competition — was once more that of an open school: three buildings separated by a copse and a lake, and connected by walkways; three buildings set on a common base but visually very different. The first was a centralized space characterized by concrete; the second, a centrifugal one characterized by metal; and the third, an incurvated maze characterized by wood, brick and glass. A real lesson in architecture, which in fact saw Portzamparc proposing that

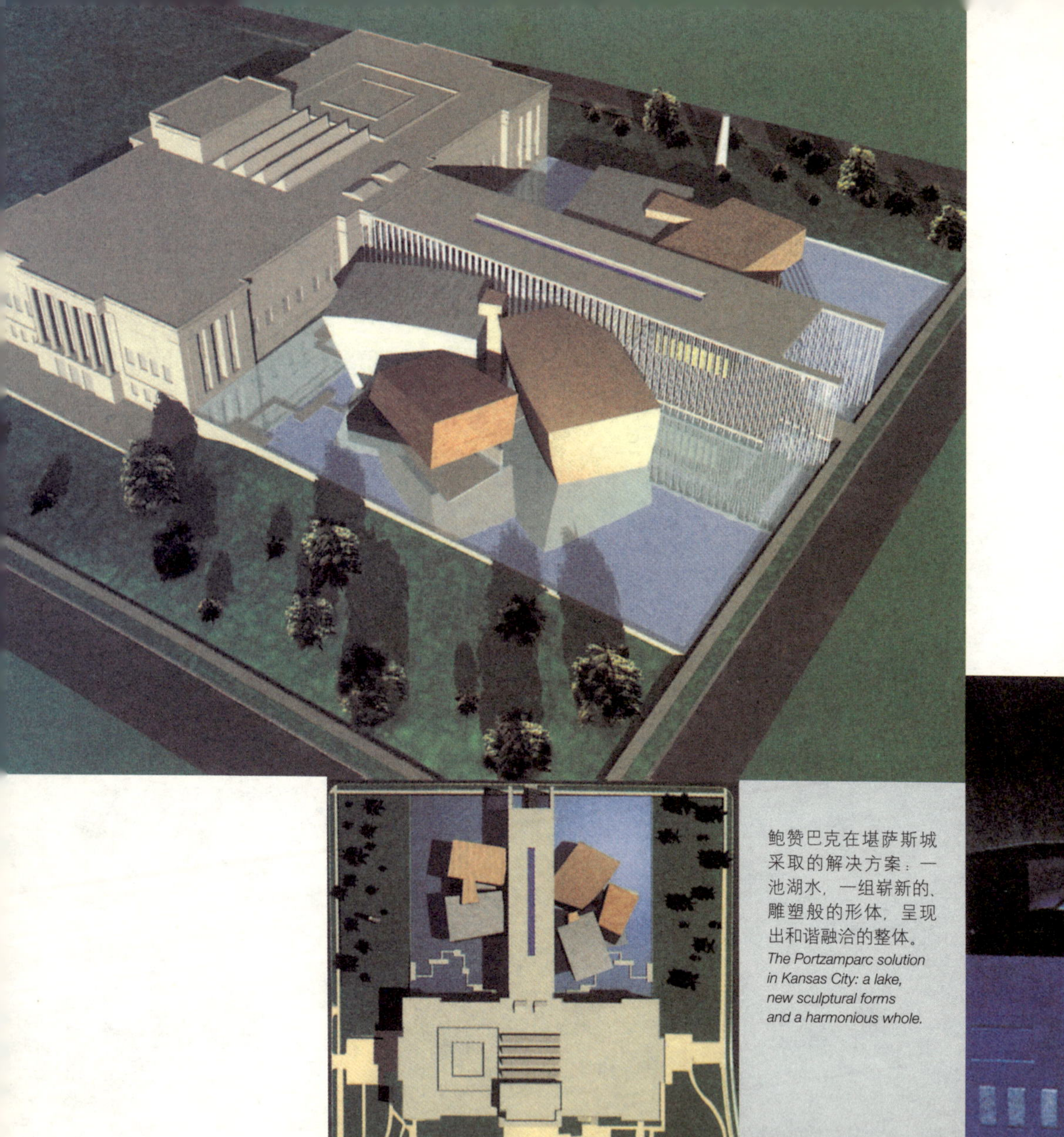

鲍赞巴克在堪萨斯城采取的解决方案：一池湖水，一组崭新的、雕塑般的形体，呈现出和谐融洽的整体。

The Portzamparc solution in Kansas City: a lake, new sculptural forms and a harmonious whole.

征出众；第三座恰如曲折离奇的迷宫，是木材、砖石与玻璃的合旋。建筑本身就是一部生动的建筑学教程，鲍赞巴克甚至提议由三位建筑师共同操刀进行设计。

1995 年的汉城，1999 年的堪萨斯城和 2000 年的蒙特利尔见证了鲍赞巴克参与的三项主要竞赛——但鲍氏均未能建功。

当时韩国美轮美奂的艺术收藏品保存在一座日本侵略者占领过的建筑内。灰色的记忆与糟糕的建筑境况激发了迁址的迫切愿望——计划迁往汉城中心的一处平原地带，美国军队曾从这里撤离，而基地中还有一条河流屈伸绵延。

鲍赞巴克着手采用韩国的传统方式设计水体：天穹回荡在水中，碧水与青天交相辉映。他堆积了四座小山来保存和储备水源，并将博物馆的四角架设在组合"桥

three architects should work together on it.

Seoul 1995, Kansas City 1999 and Montreal 2000: three major competitions — and Portzamparc lost them all.

Korea at the time kept its marvelous art collections in a building that had probably once been home to the Japanese occupation forces. Bad memories, a bad building and an urge to move elsewhere — into the heart of Seoul, to a plain the American army had just evacuated, with a river winding its way through.

Portzamparc set about using water in the traditional Korean way: as an echo of the sky. He raised four hills to shelter the reserves of water and set the four corners of his museum on the four resultant piers. Thus the museum floated over water and a garden. The underlying idea — a kind of island floating over a park and reflected in water — would reappear in the Rio de Janeiro project.

墩”之上。因而建筑俨然悬浮在水体与庭园之上。这里隐含的根本性理念——漂浮在公园上空且倒映于水中的蓬莱仙岛——在里约热内卢的项目中得以重现。

在堪萨斯城，问题立即变得完全不同，但却具有惊人的可比性。纳尔逊·阿特金斯博物馆座落在丹·凯利设计的华美上乘的公园内——这座庞大、庄严的平行六面体建筑容纳了三个收藏古典和中国艺术瑰宝的展厅——如今需要扩建："在飞往纽约的航班上，我突然领悟到，现状建筑无可挑剔的矩形形态应当保持，但我们可以利用建筑北墙面作为背景以达到这一目的。我还认识到，（现状）停车场的开放区域可以被另一完美的矩形取而代之：即一池湖水。在这稳若泰山的'平面'之上，你可以放置崭新的、雕塑般的体块。通过这种方式，旧建筑摇身变为新建筑的基石，为新建筑提供了展示的舞台，并在二者之间产生了密不可分的共荣共赢关系。重叠与和谐音：概念构思恰到好处，而成排的竖直线条延续并传承了多立克柱廊（的形态），仿佛光

In Kansas City the problem was at once totally different and astonishingly comparable. Located in a superb park designed by Dan Kiley, the Nelson Atkins Museum — an enormous, imposing parallelepiped housing three handsome collections of classical and Chinese art — was in need of extension: "In the plane on the way to New York, I suddenly realized that the perfect rectangle of the existing building had to be retained, but that this could be achieved by using the wall of the north facade as a backdrop. I realized, too, that the open area of the parking lot could be replaced with another perfect rectangle: a lake. And on this 'plateau' you could put new, sculptural volumes. In this way the old would provide a base for the new, presenting it and generating at the same time an indissociable reciprocity." Overlap and assonance, then: a remarkable concept in every respect, and one highlighted by the long gallery of vertical lines framing and containing the Doric porch intended as a light-filter.

This approach to light recurs, differently expressed, in the proposal for the Quebec Library in Montreal: "I'd designed an enormous, airy, light-filled nave protected by a shell of air-

首尔的心脏地带，一座悬挂于巨大湖泊之上的博物馆——体现了对韩国文化中天与地的交互共生关系永久性的关注。

In the heart of Seoul, a museum suspended over a vast lake – and a permanent concern with Korean culture's earth/sky reciprocity.

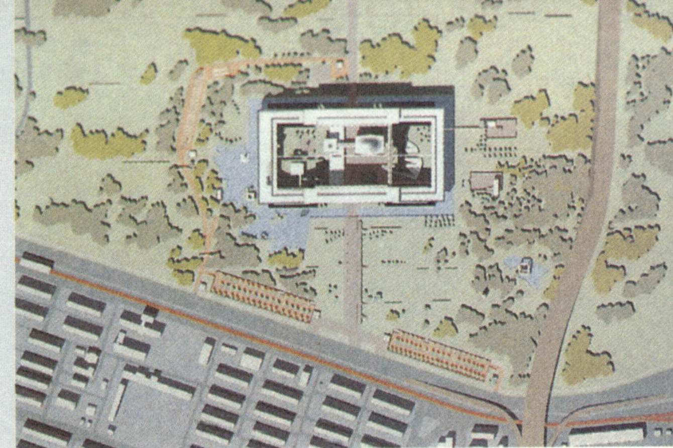

线的过滤器一般，成为设计的点睛之笔。”

类似的光影处理在蒙特利尔的魁北克图书馆设计中被诠释一新：“我希望设计一座宽敞开阔、通风良好、光线充足的中央大厅，维护结构是装有空气调节装置的玻璃幕墙，大厅空间颇为高峻，侧向光照充足，并拥有宽广辽远的城市远景视野。这座水晶宫容纳了四个灵活多变的封闭体量，划分明确、形态有机。这些场所分担任务书提及的诸多类型活动：图书馆，媒体中心，礼堂，讲座授课中心，公共设施等等。”

“夜晚，这些奇异的体型各自渲染着灯光，浮影翩跹，仿佛游弋在巨大魔术盒中的鳟鱼。”

光线的地位至高无上，而韵律与节拍也统率了底层与地下部分：毕竟，蒙特利尔一年中的生活将会有半载在地下度过。此外，建筑结构的处理方式技惊四座——设计采用三角形支撑，楼板与其他结构融为一体——这些都不禁让人联想起造船业中使用的建造方法。

conditioned glass, high enough to let in lateral lighting and offering a broad vista of the city. This 'aquarium' contained four flexible, enclosed volumes, clearly separated and almost organic. These were the venues for the various activities specified in the brief: library, media center, auditorium, lecture center, public facilities and so on.

"At night these odd silhouettes would be individually tinted with light, creating the impression of fish floating inside the big magic box."

Height for the light, then, but a tempo dictated by the ground level and the basement: after all, life in Montreal is subterranean for six months of the year. Plus a stunning handling of the structures, triangularly braced and integrated with the floors in an approach highly reminiscent of ship-building.

So the competitions came and went, generating their quota of victories, failures, disappointments, debacles, misunderstandings and misinterpretations. The most recent imbroglio for Portzamparc was the Cinema

魁北克图书馆将坐落于蒙特利尔法语区的中心地带，为城市提供亟需的焦点。

The Quebec Library was to be set in the heart of the French-speaking district of Montreal, in order to provide a badly-needed focal point for the city.

就这样，一次次的竞赛来而复去，周而复始，留下了胜利与失败，留下了失意与溃败，也留下了隔阂与误解。鲍赞巴克遭遇的最近一次纠葛是上海电影博物馆竞赛，由于截止日期迫在眉睫，他只好将蒙特利尔项目中的关键性理念悉数托出，并将体量进行破碎化处理，塑造出室内外空间的相互交融，这在加拿大自然是无法可想的。一天清晨，他带着方案模型和笔记本电脑飞往北京。目的地：长城脚下的评审会。你瞧！中国海关拒绝为他的模型通关。不必介意，陪同人员说到，您不一定要有模型。结果：到达伊始，他见到 12 件模型展示陈列在那里——自然他自己的模型不在其中。难道别人捷足先登，提前浇注了模具？不管怎样，这令人垂涎的项目最终花落一家社会关系优越的美国事务所，他们的构思说明笨拙无奇："电影院"意味着"明星"，你猜怎样，由于中美两国国旗均点缀有星形图案；因此，这庇护在长城之下的建筑的所有窗扇都将呈现星形图案……

正如鲍氏所言："当你逾越了特定的界限，一切都变得难以预料且无可挽回了。"

Museum in Shanghai, where the competition deadline was so tight that he recycled the main features from Montreal while fragmenting the volumes and creating an interior/exterior interplay that would, naturally enough, have been unthinkable in Canada. One morning we see him fly into Beijing with a model and his portable computer. Destination: a jury session not far from the Great Wall. And lo and behold, Chinese customs refuses his model a clearance. No matter, says his escort, you won't need it. Result: on arrival he sees twelve models on display — but not his own, naturally. Has the die been cast in advance?

Whatever, the coveted project goes to an already well-connected American agency, for a mind-bogglingly literal proposal: "cinema" means "stars" and guess what, our two national flags, the Chinese and the American, are both festooned with stars; so all the windows, there in the shadow of the Great Wall, are to be star-shaped...

Truly, as the man said, "Once you're over the limit, there's no drawing the line."

宽敞开阔、通风良好、光线充足的中央大厅。建筑体块覆盖了装有空气调节装置的玻璃幕墙。

An enormous, airy, light-filled nave. A block covered with a shell of air-conditioned glass..

9 登峰造极 culminating

巴黎广场朴素的立面反映出柏林重建中的种种策略法规。
The severe Pariser Platz facade reflects Berlin's reconstruction regulations.

大厅中过滤阳光的孔穴与弗朗索瓦·莫尔莱的作品相敬如宾。
The light-filtering cavities in the lobby dialogue with the work by François Morellet

柏林法国驻德大使馆

The French Embassy in Berlin

音乐城落成 8 年后，2003 年 1 月 23 日，鲍赞巴克设计的新法国大使馆在柏林落成。

很多人认为音乐城已然堪称举世无双的成就，但它的创造者却只当其为学艺历程中的跬步而已。如今，这新一轮的实践又被看成登峰造极之作：功德圆满的集合体与精湛技艺的全面展示。

建筑师对些这论断仍旧不以为然，不仅反唇相讥，而且还将他接下来的（许多）设计方案作为引证：种种新的历险与探索（逐渐）浮出水面，在卢森堡和里约，他投身全然未知的领域，涉猎新的主题、设计模式与表达方法。

那么，（让我们）来到柏林。自从（人们）在地图上勾画出一条分界线开始（代指柏林墙——译者注），柏林就摇身变成了最为奇特怪异的城市。它包括一系列的村庄、湖泊与森林，最初是勃兰登堡选帝侯时代的首府，后来又先后成为普鲁士、德意志帝国、第三帝国、纷争时期（指东德、

On 23 January 2003, eight years after completing the Cité de la Musique, Portzamparc handed over the new French Embassy in Berlin.

For many the Cité had been a crowning achievement, yet for its creator it was no more than a step in the learning of his craft. This time round, the stunning new exercise was seen as a culmination: a consummate synthesis and a demonstration of total mastery.

And once again the architect was in open disagreement with these judgements, both verbally and in terms of subsequent projects: for other adventures were on the horizon, the Luxembourgs and the Rios that would see him on fresh territory, working through fresh themes, solutions and modes of expression.

Berlin, then. Strangest of cities ever since it was created with a line drawn on a map. A succession of villages, lakes and forests, initially seat of the Electors of Brandenburg then capital, successively, of Prussia, the German Empire, the Reich, the time of discord and, finally, Germany. Which Berlin are we to believe in? The shady, delectably lascivious creation of Alfred Döblin, or

西德分裂时期——译者注）的首都，最后则成为德国首都。我们应当信奉哪一个柏林？是艾尔弗雷德·德布林作品中那阴郁灰暗、情欲勃发的城市，还是约翰·勒·卡雷笔下描述的单调乏味，却如幽灵般险象环生的世界？是巴布斯特还是法斯宾德的城邦？1989 年，这座城市一举惊天：柏林墙轰然倒塌，数英里长的铁丝网、围栅、刺栏也随之悉数消失。

具有悖论意味的是，正是柏林墙的拆除使得这座城市荣升为世界建筑业的首善之区，凭借其出众的品质令上海退居次席。大队人马（随之）慕名而来：（他们大都是）在国际建筑界享有盛誉、被各类大型设计竞赛争相邀请的明星建筑师。

英国人诺曼·福斯特为国会大厦装上了崭新的穹顶，一座硕大、且周边镶嵌有钢制条纹的锃光瓦亮的圆葱形屋顶。意大利的伦佐·皮亚诺承担波茨坦广场的设计，将柏林的现实情境，巴黎的城市气质，以及意大利的氛围融为一体。距此一箭之遥，则是英国人理查德·罗杰斯和德籍美裔建筑师海莫特·扬设计的高技派表现主义作品。而法国人让·努韦尔在拉法耶特拱廊项目中运用了（玻璃）圆锥体造型和丝网印刷玻璃。以色列的丹尼尔·里伯斯金将犹太教的六芒星形打散分裂，创造出犹太人纪念馆的平面形态。德国人约瑟夫·克莱胡斯将汉堡火车站改建为现代艺术博物馆。法国人克劳德·瓦斯克尼对遗留的波希机车厂进行了再开发。在东边很远的地方，另一位法国人多米尼克·佩罗建造了一座魔咒般不可思议的体育综合体——室内赛车场和奥林匹克游泳中心——它们周围还环绕着从诺曼底

通往二层的楼梯，以及尼尔·托罗尼的作品。
The staircase to the second floor, with the work by Niele Toroni.

the drab, spookily perilous one of John Le Carré? The city of Pabst or the city of Fassbinder?

In 1989, the bombshell: the Wall collapsed and with it disappeared miles of barbed wire, fences and spiked chevaux de frise.

In a paradox to beat them all, it was the fall of the wall that rapidly made Berlin the world's architectural capital, using quality to beat Shanghai into second place. This was when new regiments in black hit town: the international architectural celebrities whose names are at the top of the list whenever a big competition is announced.

So the Reichstag was newly topped with a gigantic, incredibly light cupola, an onion dome wrapped around with steel striations by Englishman Norman Foster. The Postdamer Platz was taken in hand by the Italian Renzo Piano, mingling Berlin reality, a Paris feel and impressions of Italy. A stone's throw away Englishman Richard Rogers and Germano-American Helmut Jahn threw themselves into Expressionist high-tech, while France's Jean Nouvel stood cones on their heads and screenprinted glass for Galeries Lafayette and Israel's Daniel Liebeskind shattered the Star of David to create the floor plan for his Jewish Museum. German Josef Kleihus turned the old Hamburg Station into a museum of contemporary art, Frenchman Claude Vasconi redeveloped the former Borzig locomotive plant. Far, far away to the east, another Frenchman, Dominique Perrault built a weird, mysterious sports complex — velodrome and Olympic swimming pool — ringed with apple trees from Normandy. Closer, however — much closer — and directly opposite the site for the French Embassy, that most Angelino of Canadians, Frank Gehry, brought his spectacularly fractured style to bear on the headquarters of the DG bank.

移植来的苹果树。在近一些的地方——（实际上）近得多——法国大使馆基地的正对面，最具洛杉矶气质的加拿大人富兰克·盖里在DG银行总部设计中沿用了自己蔚为壮观的断裂（解构）风格。

鲍赞巴克就是在这样浓郁的氛围中埋头设计自己的大使馆的。他去柏林踏勘，感知和体恤那片局促狭窄的基地，并开始默想自己敬仰的建筑师汉斯·夏隆，这位德国人"精心致力于每一项任务，仿佛每次设计都是自己的处子演出"。从很大程度上说，这也是惟一的选择。因为（现状）场地十分促狭——约140米进深，并被国会办公建筑一堵无开窗的28米高墙所限定，几乎变身为一座"混凝土浴缸"（弗朗西斯·拉姆伯特，《费加罗日报》；Francis Rambert，Le Figaro，03.10.2003）。

在柏林，鲍赞巴克游历并朝拜了夏隆的爱乐音乐厅，这雄浑而错综复杂的建筑力作曾激励法国建筑师踏上自己的长征之路。他凝视这座声效至上的建筑，回想起夏隆的另一段评论："音乐厅的形状源于地形景观的启示，你可以称这里的空间为山涧河谷：最远端是演奏乐队，藤蔓覆盖的斜坡沿周边升起。与这地景搭配附和的是那巧夺天工的天花板。声音不应从大厅的某个角落发出，它将从中央款款倾泻而出，从四面八方向听众们袭来。"

鲍赞巴克会心一笑，联想起音乐城、奈良、卢森堡、里约和其他设计，深感自己已然领悟到夏隆设计思想的真谛，同时又能不困囿其中。

回到巴黎，他埋头苦干，最终击败亨利·高丹和让·努韦尔，并赢得了竞赛。

勃兰登堡门茕茕孑立、若隐若现，成为记忆与象征，自1865年柏林的第一面城墙被拆除开始，它便一贯如是。如今，它较任何时代都更为与众不同——形态孤立、不具功能性、满怀僧侣气质：它不再是一扇简简单单的大门，却宛若一座纪念

极度局促的基地好像"混凝土浴缸"一般，建筑师技高一筹，布置了7座相互关联的独立建筑。

The extremely narrow plot had become a "concrete bathtub" into which, with consummate skill, the architect slipped seven distinct but interrelated buildings.

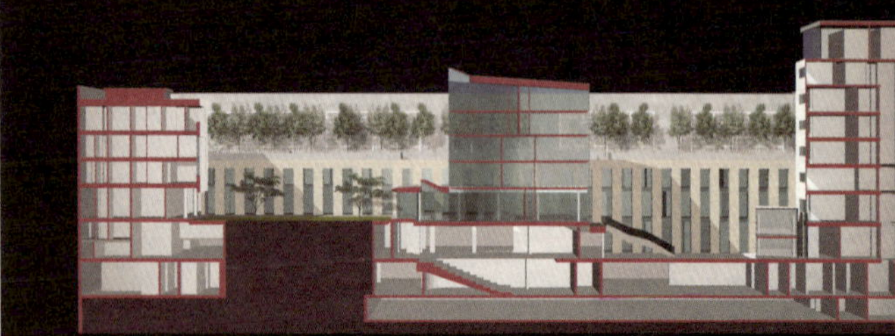

It was in this hothouse atmosphere that Portzamparc worked away at his Embassy project. He went to Berlin, felt out the odd plot the Embassy was to be built on, began thinking about his much-admired Hans Scharoun and remembered that the German architect liked to "go into each specific task as if it were the first time". This was pretty much the only possibility given the narrowness of the plot, 140 meters deep and transformed into a "concrete bathtub" (Francis Rambert, *Le Figaro*, 03.10.2003) by Reichstag office buildings offering a long, windowless wall 28 meters high.

Being in Berlin, Portzamparc undertook his pilgrimage to Scharoun's Philharmonic Hall, the boldly complex work that had triggered the French architect's long march. As he contemplated the building in which sound was all, he recalled another remark by Scharoun: "The shape of the concert hall is inspired by landscape. You could call the space a valley: at the far end is the orchestra, with vine-covered slopes rising around it. Matching this earthly landscape is the celestial landscape of the ceiling. The sound must not emerge from some corner of the hall, but from its middle and its depths, descending on the audience from all sides."

Portzamparc smiled to himself, recalling the Cité de la Musique, Nara, Luxembourg, Rio and the others, and feeling he had absorbed Scharoun's lesson well, yet freed himself from it.

Back in Paris, he plunged headlong into the competition — and won, against Henri Gaudin and Jean Nouvel.

The Brandenburg Gate loomed large for him, as memory and symbol, standing alone as it had since the leveling of the first wall around Berlin in 1865. Isolated, function-free, hieratic, it was more intensely present than ever: no longer a mere gateway, but a monument in its own right. For the moment, though, it was Karl Friedrich Schinkel he had on his mind. Schinkel who had imposed his romantically neoclassical vision on the city in the early 19th century, bequeathing it a set of rules that still stands.

碑。此时此刻，鲍赞巴克脑海里又闪现出卡尔·弗里德里希·申克尔的身影。在19世纪早期，申克尔为这座城市注入了浪漫的新古典主义风韵，将一系列至今仍被尊行的准则与惯例馈赠给柏林。

申克尔与勃兰登堡门：他们如影随形。在1945年对柏林的轰炸中，勃兰登堡门奇迹般地逃过一劫，但其周边地带却在随后的50年时间里片瓦无存、杳无人烟。同样位于巴黎广场的旧法国大使馆被夷为平地，仅遗存两尊雕塑——一对青铜猎犬和一名端坐的男孩雕像——鲍赞巴克将它们遴选出来，为自己的新建筑增光添色。

因此，旧巴黎广场和名声显赫的菩提树下大街的早期遗迹已经荡然无存，惟有勃兰登堡门风采依旧，自成为声名狼藉的查理检查站（Checkpoint Charlie，东西德之间的关卡——译者注）开始，它就恢复了建成初期的关卡和通行道口身份。

鲍赞巴克构思的面向广场的立面“朴素却又亲切”（《每日镜报》，Tagesspiegel，01.30.2003），与柏林的“批判性重建”策略珠联璧合。建筑师游刃有余地运用申克尔的设计手法处理基座、檐口与阁楼，严谨而率直地对待开窗：不对称的楔状窗洞在外部给人以活力旺盛之感，在内部则形成别致优雅、明亮适度的壁龛。

进入大使馆的最佳路线是取道威廉大街，并沿绵长的室内街道步行前进。这条街巷6米宽，空间高耸眩目，用产自巴黎的卵石铺砌，伫立两侧的混凝土墙面镶嵌着条纹肌理：整体效果无可辩驳地怀有巴黎气质，但更加奇异的是，它甚至让人神游在文艺复兴前的意大利。在这里，“激励”的手法同样耳熟能详，来访者愈走愈能感觉到大使馆展示了一处生机勃勃的室内生活场所。（建筑的）启示应运而生：一组互不雷同、散布丛生的建筑群沐浴着和煦灿烂的阳光，仿佛一把羊踞骨恶作剧般地随意投掷到盒子中。默斯·康宁安的精魂

精明地处理好连接部件、透明性与视觉效果绝非易事。

It was no easy task dealing with linking sections, transparency and views.

Schinkel, Brandenburg: no getting away from them. By some kind of miracle the bombing of Berlin in 1945 had spared the Gate, while the surrounding area was to remain a totally devastated no man's land for over fifty years. The previous French Embassy, also situated on the Pariser Platz was obliterated, except for two sculptures — a pair of bronze dogs and a seated boy in stone — that Portzamparc co-opted for his new building.

Thus there remained absolutely nothing of the Pariser Platz or the beginning of the venerable Unter Den Linden ("Under the Lime Trees") except the Brandenburg Gate, which had regained its former status as barrier and crossing point when it became the notorious Checkpoint Charlie.

The facade Portzamparc offered the Pariser Platz was "severe but warm-hearted" (*Tagesspiegel*, 01.30.2003), a perfect fit with Berlin's "critical rebuilding" strategy and regulations. The architect allowed himself a free hand with Schinkel's rules for the base, cornices and attic and the strict, straightforward treatment of the windows: the latter's dissymmetrical embrasures create a sense of energy on the exterior, while inside they become elegantly appropriate light-filled niches.

However, the best way to approach the embassy is from the Wilhelmstrasse, taking the long indoor street. Six meters wide and almost dizzyingly vertical, it is paved with Paris cobblestones and lined with a long, high wall of striated textured concrete: the overall effect is irresistibly reminiscent of Paris and, more curiously, of a pre-renascent Italy. The "aspiration" technique is in evidence here too, and as the visitor advances the embassy begins to reveal

徘徊在大使馆中，人们仿佛进行了一次城市漫游。
A walk through the embassy is like a stroll through town.

再次跃然纸上。"一切随缘而定，"编舞艺术家曾坦言，他驳斥并拒绝——在1950年代前期——传统的循序渐进式的结构概念，甚至是基本的理性序列观念。这一编舞的新途径运用随机性和偶然性，并找到了合乎逻辑的表达形式，当时，康宁安义无反顾地采用投掷硬币的方式来决定（舞蹈动作的）运动方向、速度和频率，也借此编排组织舞蹈的步伐。

建筑、羊跖骨、硬币：游戏的冠名或许是"偶然性"，但很显然鲍赞巴克理解、控制、并——尤其可贵的是——超越了它。康宁安是值得仰仗的，但同样会有新的趋势诞生，想一想泰拉·萨普的作品，他从古典主义艺术中汲取了无与伦比的清晰性与精确性；再想想弗雷德·阿斯泰尔那无与匹敌的典雅与别致；还有迪斯科舞的兴高采烈气氛。

所有这一切完美地交织，使鲍赞巴克更上层楼，领悟到卡罗尔·阿米塔奇那具有冒险气质的断奏曲和痉挛似的节奏韵律。舞蹈成为空间重构的金钥匙。但同时雕塑性也若隐若现，他如同亨利·摩尔和亚历山大·考尔德那样探索虚空与穴隙的世界，并将极少的物质与海量的空间协调组织，建构起二者间的关系。

谈及虚空与盈满的构建，我们无法忘记电影："电影像手风琴一样处理时间：点滴的时间可以被延展放大——使你在短短三个镜头中历经十载。"

眼神深邃豁达，遥及目光穷尽之处：这就是建筑师在柏林希求的东西。一组若即若离的建筑群与庭园景观穿插交织，精妙绝顶，庭园结合了青草、中国白桦、常春藤、蛇麻草、灌木、以及一座垂直的紫藤园，设计均出自雷吉斯·吉纳德之手。

当来访者继续着名副其实的"城市漫步"，显而易见，景观、场景与表达方式的多样性与丰富性成为鲍赞巴克建筑风格的精粹：它如同文学一样，将激发各种情感作为第一要务。

犹为可贵的是，垂直线条与水平线条往来交错，掘进伸展、膨胀扩散，逾越种种桎梏，产生出场所的表象与深度，流露出动人心魄的诗意与精湛技巧。最终成果依循展露、深景、突进、攀升等手法与形

itself as a place with a vigorous internal life. Then comes the revelation: in the full light of day, a scattering of different buildings looking like a handful of knucklebones mischievously tossed into a box. And once again there emerges the guiding spirit of Merce Cunningham. "Everything follows anything," the choreographer once said, refuting and rejecting — in the early 1950s — the traditional concept of a structure whose events gradually reach a conclusion, and even the basic notion of rational sequence. This new approach to the structure of dance was to find logical expression in the use of chance, with Cunningham not hesitating at the time to throw a handful of coins on the ground as a way of fixing the direction, speed and frequency of the movement, and the composition of the steps.

冗长无趣的墙面因一排白桦树增色良多;空间延伸的能力激发了一系列庭园的创造。

A long, blank wall canceled out by a line of birches; and a capacity to expand space that allowed for the creation of several gardens.

式，造就无穷无尽、出其不意的闪光点，与尼采提及的那种“与混乱无序格格不入的令人敬畏的多样性”如出一辙。当我们游走在大使馆里，会不由自主地惊诧于这座通廊（或称“浴室”，如果你愿意的话）沐浴在光影世界中的翩翩倩影——那是晨曦中的第一缕霞光。种植齐整的接待区域依然如是，它毅然植根于大地沃土，绝非先入为主的空中花园印象可资比拟。

伊丽莎白的室内设计——她与丈夫配合默契，共同赢得了竞赛——与建筑设计的气质一脉相承。尤其值得一提的是艺术品的遴选，涉足的艺术家有弗朗索瓦·莫尔莱、乔治·诺尔弗朗索瓦·雷昂、尼尔·托罗尼、马丁·华莱士，甚至还包括赵无极。

同样锋芒毕露的还有吉尔·阿尤、皮埃尔·亚勒金斯基、莫尼克·弗里德曼、汉斯·哈同、伊夫·克莱因、琼·米切尔、克劳德·维亚拉的作品，它们是从国家现代艺术基金会和国家家具管理委员会借调的艺术佳作。无可否认，建筑的流线有些离奇怪诞，8字形态象征了盘根错节与广

Buildings, knucklebones, coins: chance might be the name of the game, but it was clear that Portzamparc understood, controlled and — especially — outstripped it. Cunningham was the anchor, but there were new movements too, reflecting the work of Twyla Tharp, who borrowed from the classics with such clarity and precision; the unparalleled elegance of Fred Astaire; and the joie de vivre of disco.

All these came together perfectly, taking Portzamparc further still, towards the perilously staccato, convulsive rhythms of Karole Armitage. This was dance as a key to spatial restructuring. But there was sculpture here too, for digging into the void like Henry Moore or Alexander Calder, for organizing the relationship between a very small quantity of matter and enormous quantities of emptiness.

Not forgetting, of course, the cinema, for structuring fullness and emptiness: "The cinema manipulates time like an accordion: a single minute can loom very large — and then you cover ten years in three shots."

Magnify, dilate, cast the eye into the furthest distance: this is what the architect is talking about in Berlin. The result being that scattering

各式各样的材料与肌理制造了节奏韵律、蓬勃向上的态势和更广阔的视野——简而言之，一次真正的旅行。

The variety of materials and textures generates a rhythm, an upward thrust, a broader view - in short, a real itinerary.

垂直线条与水平线条往来交错，创造出无穷无尽的惊喜，探究、逾越并拓展了建筑的种种阈限。

No end to the surprises triggered by the interplay of vertical and horizontal lines that cut in, push out and extend the architectural limits.

袤无垠——也使人联想起纽约路易威登大厦那隐喻随机性的 7 字形立面，和奈良的莫比乌斯带。在柏林，法国大使馆拥有一系列的创新与惊喜，堪称情感与知觉的世界，因此毋庸置疑：它是登峰造极的成就和无可争辩、功高盖世的杰作。

在这里，鲍赞巴克实现了马蒂斯的格言，“我笔耕数载，为的就是让人们能够有感而发，‘马蒂斯——也不过如此而已’！”

延伸自威廉大街的“通廊街巷”统辖了建筑的整体，蕴涵着巴黎与意大利的气质。其尽头则是咖啡厅。

The long through street from the Wilhelmstrasse unifies the whole with suggestions of Paris and Italy. At the far end: the cafeteria.

of buildings whose intervening landscapes are orchestrated with infinite subtlety by Régis Guignard's mix of grasses, white Chinese birch, ivy, hops, bushes and a vertical garden of wisteria.

As the visitor continues with what can quite aptly be termed a "stroll through town", it becomes clear that diversity and multiplicity — of landscape, situation and mode of expression — are the quintessence of Christian de Portzamparc's architectural style: a style which, in the same way as literature, puts its primary emphasis on the power to evoke.

This notably includes an interplay of vertical and horizontal lines that hollow out, expand, dilate, push back limits and generate maskings and depths of field whose poetry and virtuosity are nothing short of astounding. The outcome takes the form of revelations, vistas, plungings and soarings, an endless string of surprises deserving of Nietzche's reference to "A formidable multiplicity at the other extreme from chaos".

As we move on through the embassy, we are amazed at the way this corridor ("bathroom", if you like) is drenched in light — the light of early morning, what's more. The same is true of the planted-out reception area, not the hanging garden it first appears, but one rooted firmly in the ground.

Elizabeth de Portzamparc's interiors — prepared in association with her husband; they were joint winners of the competition — are in the same vein. Especially worthy of note is the choice of art works, the list including François Morellet, Georges Noël, François Rouan, Niele Toroni, Martin Wallace and even Zao Wou-Ki.

Also in evidence are works by Gilles Aillaud, Pierre Alechinsky, Monique Frydman, Hans Hartung, Yves Klein, Joan Mitchell and Claude Viallat, lent by the FNAC (National Contemporary Art Collection) and Mobilier National (National Heritage). The itinerary has its own undeniable strangeness, being in the form of an 8, the symbol of intertwining and infinity — but also a reminder of the LVMH tower facade in New York, with its 7 symbolizing chance, and of the Moebius strip in Nara. In Berlin, then, the French Embassy, with its succession of new developments, surprises, physical sensations and emotions, leaves no choice: this is a masterpiece and as such, a true culmination.

This is the fulfillment of a Matisse saying, "I've worked for years to make people say, 'Matisse — that's all there is to it'!"

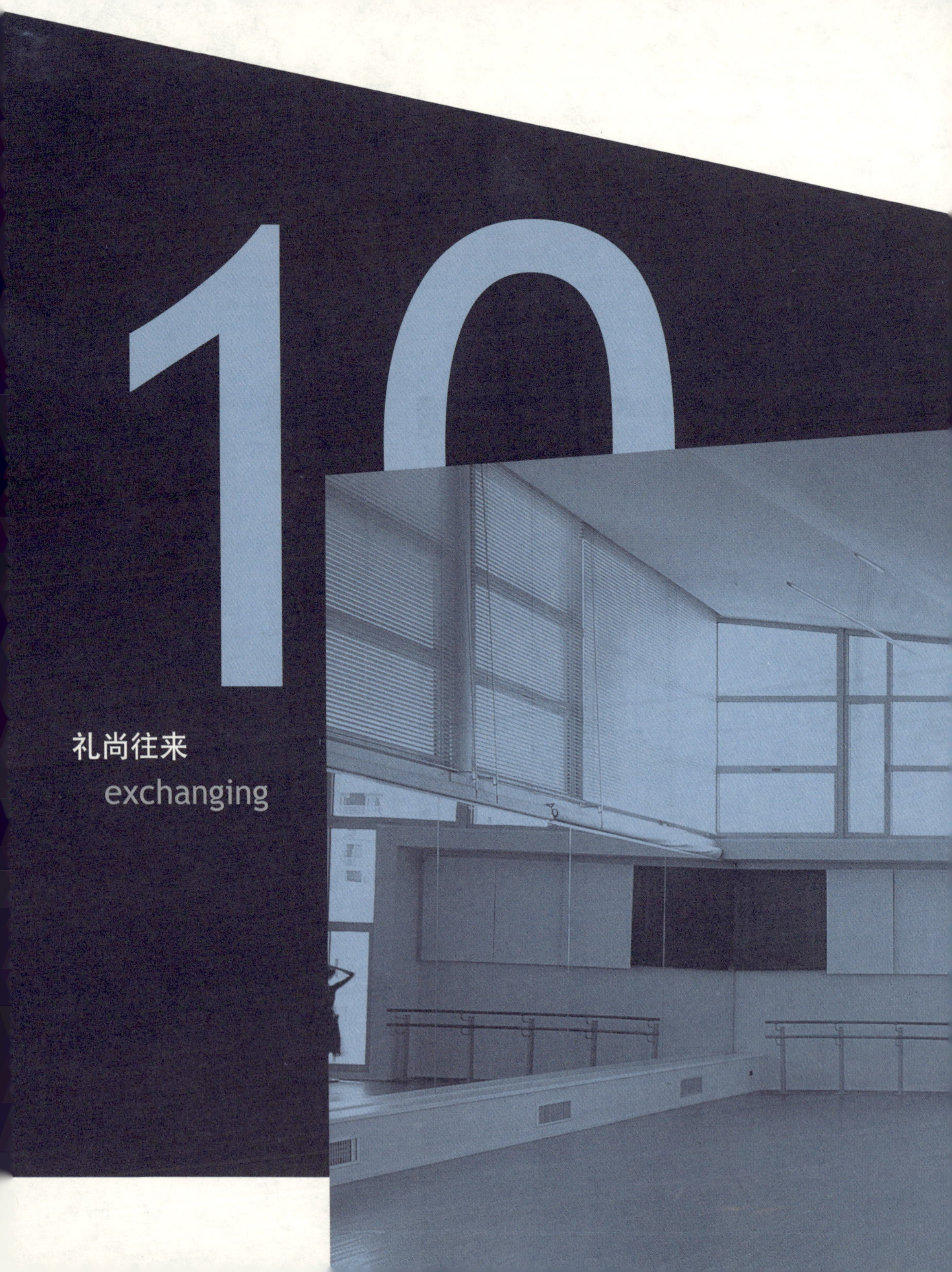

10

礼尚往来

exchanging

创造性的舍与得

Creative give and take

柏林的法国大使馆来之不易，鲍赞巴克与妻子——也是共获殊荣的竞赛合作伙伴——伊丽莎白共创辉煌：他们使用法国政府为公共建筑中艺术品陈设提供的资金补贴——在建筑总投资的1%——收集到弗朗索瓦·莫尔莱、乔治·诺尔、尼尔·托罗尼、马丁·华莱士和赵无极等艺术家的作品。

当你从巴黎广场踱步进入建筑，莫尔莱、托罗尼和马丁·华莱士的作品映入眼帘。莫尔莱的设计形如马赛克，隐约勾勒出拉维那的几许回忆，它特性鲜明，在光线被奉若神明的空间中"熠熠生辉"。这是恰到好处、直觉至上的选择，倾诉着一

No mean feat, the one pulled off at the French Embassy in Berlin by Christian de Portzamparc and his wife and fellow-laureate Elizabeth: with the subsidy offered by the French State for art in public buildings — 1% of the total cost of the building in question — they managed to include works by François Morellet, Georges Noël, Niele Toroni, Martin Wallace et Zao Wou-Ki. As you step in off the Pariser Platz, Morellet, Toroni and Wallace are right there. Morellet's contribution is a kind of mosaic, a distant reminder of Ravenna whose particularity is to "shine" in a space where light is at a premium. A truly appropriate, highly intuitive choice that tells its own story behind of a couple united, separated then back together again. Since Morellet's intelligence, wit and humor are part

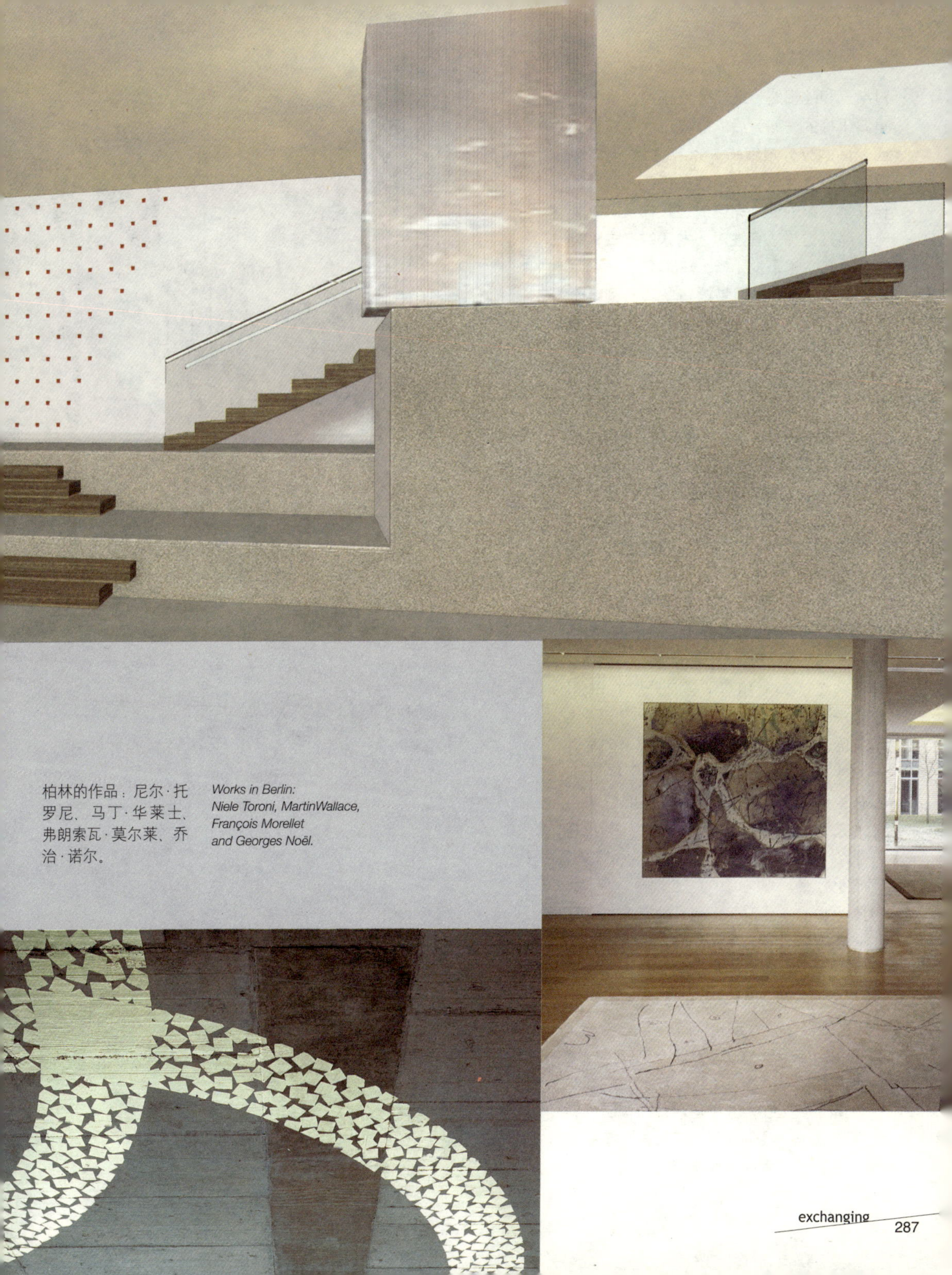

柏林的作品：尼尔·托罗尼、马丁·华莱士、弗朗索瓦·莫尔莱、乔治·诺尔。

Works in Berlin: Niele Toroni, MartinWallace, François Morellet and Georges Noël.

对早先和睦相处、继而分道扬镳、最终又重归于好的夫妇背后的故事（代指东德与西德间的分分嚷嚷——译者注）。莫尔莱的才情、睿智与幽默风趣久负盛名，因此作品体现出的近乎完美的适宜性便不足为奇，而且这种适宜性对艺术家的自由精神与独立自主也毫发无损。1986 年，多米尼克·博索在尚贝里博物馆写就《马列维奇的幽灵》时提到，"莫尔莱的作品拒绝纯粹而简单的完整性和毗邻并置状态，因此保留了自足性与竞争的能力。更为精明的是，他将绘画 / 雕塑与建筑之间通常存在的关系加以变更，而且不出意料，其作品还以局部环境背景的展现为形式，令人们产生反常规的基本理解。"还有什么语言能更确切地描述他的柏林作品中彻头彻尾的适宜性、针对性和几近鲁莽武断的气质呢？

托罗尼的作品同样绘制在墙面上，与莫尔莱的作品对照呼应。在四剑客当中——布伦，莫塞，帕芒蒂埃，托罗尼，他们曾在 1966 年组建了昙花一现的 BMPT 小组——托罗尼是唯一始终迷恋于徒手画笔绘画的艺术家，其他人则埋头研究垂直与水平条带或圆形创作模式。历经近 40 载的磨砺，托罗尼体现出的激进特征与艺术姿态仍旧光亮如新、毫不褪色："我将一切创作者的主观性影响剔出绘画之外，因而也杜绝了观瞻者的主观影响。"不过，他在柏林的作品韵律十足，拥有令人敬慕的音乐感。

在两件杰作之间（代指莫尔莱与托罗尼的作品——译者注），在楼梯的角落中，马丁·华莱士——他是一位在布列塔尼工作的英国雕塑家和版画家——创作的纪念性雕塑被束之高墙之上：一块墩实的棱柱发光体，积压在两片甲基丙烯酸树脂材料之间。这里，我们感到了谦逊与自身的消隐，佐证了艺术家与建筑师之间长久而精

格拉斯的矛状（装饰），与马丁·华莱士合作设计。
The shafts in Grasse, designed in collaboration with Martin Wallace.

奥蕾莉·内穆尔，《毫米节拍》。
Aurélie Nemours, «The rythmn of millimeters».

of his reputation, there is nothing surprising about this perfect appropriateness, which in no way excludes the artist's freedom and autonomy. As Dominique Bozo wrote in 1986 of the Ghost of Malevich in the Chambéry Museum, "Here Morellet rejects both integration and juxtaposition pure and simple for his work, which thus retains its autonomy and capacity to compete. His subtlety extends to modification of the usual relationship between painting/ sculpture and architecture, adding, expectedly, a reversal of the fundamental reading of his art in the form of the partial presence of the context." What more could be said about the sheer rightness, pertinence and impertinence of his Berlin work?

Toroni's contribution is on a wall, too, but painted and in dialog with the Morellet piece. Of the four musketeers — Buren, Mosset, Parmentier, Toroni — who founded the instantly dissolved BMPT group in 1966, Toroni was the only one to stick to the unaided trace of the brush, leaving the others to get on with their vertical bands, horizontal bands and circles. Almost forty years down the track the radical character of his stance is as clear as ever: "To rid painting of all possibility of subjective intervention by the painter, and thus by the spectator." Nonetheless, the scansion of his Berlin offering has an admirable musicality to it.

Halfway between the two and halfway up the wall, in the angle of a staircase, is a luminously monumental sculpture by Martin Wallace, an English sculptor and printmaker working in Brittany: a chunk of light, a prism squeezed between two sheets of metacrylate. Here we sense a modesty, a self-effacingness that testifies to a long, spirited complicity between artist and architect: Martin Wallace has often contributed to Portzamparc's work, usually — such is the closeness of their relationship — as co-designer, co-writer. One of these joint ventures was the colonnade in Grasse, that alignment of assegais, papyrus leaves and standing oars that achieves such handsome effects of light and shade. Another was the ornate red concrete facade in Rennes, with its series of obliques and it surface in relief. In the main reception room a splendid carpet by Georges Noël is complemented by two of his canvases in a similar vein. The 1% subsidy, then, was used with impressive rigor and accuracy — and yet, says Portzamparc, this task is not always an easy one. The place, the timing, the realities of the situation are not

神饱满的合作。马丁·华莱士时常会助鲍赞巴克一臂之力，他们的关系亲密无间，经常合作设计或共同写作。格拉斯的柱廊是他们同舟共济的实例之一。作品仿佛是一排严整的长矛或纸莎草叶，又好似挺立的船橹，营造出慷慨大方的光影效果。另一成果便是雷恩项目中那绚烂多姿的红色混凝土立面，它附着有密密层层的倾斜线条，表面凹凸不平，宛若浮雕。柏林大使馆的主接待厅里陈列着乔治·诺尔设计的无与伦比的地毯，它也衬托起乔治创作的另外两幅拥有类似纹理效果的油画。1%的补贴款被严密而精确地使用——然而，鲍赞巴克承认，这并不总是唾手可得的成果。时间、地点和现实情况并不总像这里一般得天独厚，并且几个世纪以来，艺术家与建筑师的立场改弦易辙，他们的关系今非昔比，变得更为紧张了。然而这并未阻碍鲍赞巴克频繁地诚邀艺术家参与工作——他们是设计的发起者而非跟风从众的角色。艺术家的成果始终倾向于“音乐的”提炼与抽象，恰如贝阿特丽丝·卡萨德修与奥蕾莉·内穆尔在楠泰尔和音乐学校演绎的那样，又如皮埃尔·毕赫葛里欧在楠泰尔、音乐学校、和圣特·克莱尔教堂中的贡献那样。对卡萨德修和内穆尔来

在楠泰尔的舞蹈学校设计，鲍赞巴克邀请贝阿特丽丝·卡萨德修和皮埃尔·毕赫葛里欧参加。
For the dance school at Nanterre, Portzamparc called on Béatrice Casadesus and Pierre Buraglio.

毕赫葛里欧再露锋芒，这次是在拉维莱特音乐城和圣特·克莱尔教堂。
Buraglio again, this time at the La Villette Conservatory and the church of Sainte Claire.

在萨伯特教堂举办的“活的历史”展览中设计的装置草图：鲍赞巴克与毕赫葛里欧同盟的有力见证。
The preparatory drawing for the "Living History" installation at the Salpêtrière: further proof of the entente between Portzamparc and Buraglio.

always as favorable as they were here, and over the centuries the architect/artist relationship has changed, shifted its ground, become more tense. This has not prevented him, however, from calling on artists frequently — as sponsor rather than sponsored. Always there has been a clear inclination towards "musical" abstraction, as illustrated by the presence of Béatrice Casadesus at Nanterre, Aurélie Nemours at the Conservatory and, especially, Pierre Buraglio at Nanterre, at the Conservatory and at the church of Sainte Claire. As with Casadesus and Nemours, at Nanterre the issues are rhythm, musicality, tension and scansion. And most often the attention focused on acoustic paneling, thus rendered alive, vibrant, vibratile.

At Sainte Claire's, right at the top of the Avenue Jean Jaurès and restored while the Cité de la Musique was going ahead, the question was one of qualitative upgrading: creation of a portal worthy of the name and a cross that would be both symbol and signal. What Portzamparc likes about Buraglio is his way of starting out from nothing — or almost — and then transforming and transfiguring with infinite delicacy: Della Robbia-style insets, you might say. Here Buraglio would play with lacquered or enameled sheet steel : ultramarine for the portal and a brown reminiscent of the habit of the Poor Clares for the sign and part of the cross.

The complicity, the friendship between Buraglio and Portzamparc went back a long way: to

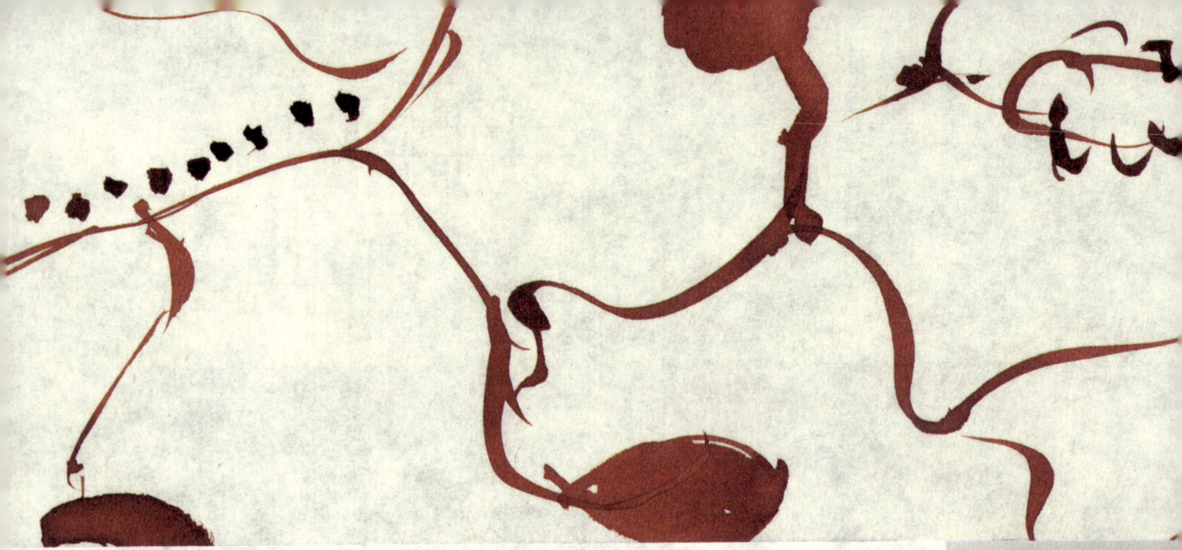

讲，楠泰尔的创作关键是韵律、乐感、张力与节奏。吸音板的设计是重中之重，它们形态鲜活、颇具脉动振颤效果。

圣特·克莱尔教堂位于让·饶勒斯大街上，其修缮工作与音乐城的建设同期进行，项目的第一要务是完成品质卓越的改良与升级：精雕细刻出无负其盛名的入口大门和兼备象征性与标志性的十字架。鲍赞巴克十分赞赏毕赫葛里欧从零做起的设计方式——或几乎白手起家——而他后续的方案深入与改善工作同样令人拍手称奇：你甚至觉得那是德拉·罗比亚风格的艺术品。毕赫葛里欧在这里演绎并运用了喷漆和瓷釉彩饰钢板：群青色大门映衬着局部呈棕褐色的十字架与标志牌，不禁使人联想起嘉兰修女会的陈规旧习。

鲍赞巴克与毕赫葛里欧的同道之情和弥笃友谊由来已久：他们在艺术学院读书时便交往甚密，还曾在1968年共谋职事。但除却彼此间的忠诚之外——或曰位居其上——如果将他们的作品比肩相较，便会觉察到二者的设计风格确定无疑地蕴含无可辩驳的一致性。作品的形制也许迥然相异，但毕赫葛里欧在圣特·克莱尔的贡献与鲍赞巴克在“新趋势”（Strada Novissima）展览中的设计貌似无关，实则神魂聚合，尤为可叹。

鲍赞巴克的另一位知音是雅克·马丁内斯，他曾在格拉斯法院的中厅内设计了60余米长的檐壁：这项作品高高在上，激励来访者“高举头颅”，体现出艺术家狡黠敏锐的直觉。马丁内斯的言论诠释了美国建筑师罗伯特·文丘里的设计思想，“附加之物（指装饰——译者注）并不削损建筑”。1990年，在进行格拉斯项目之前8年，查威尔·吉拉德在Jau城堡举行的马丁内斯画展上写道，“这些绘画使我联想起鲍赞巴克在巴黎设计的欧风路住宅，满怀悖谬气质的罅隙和开口环绕在高大的中心轴近旁……马丁内斯反复运用建筑师称为‘过度邻近’的手法。”格拉斯的作品略显冷淡，有时甚至歪扭笨拙，但它柔嫩动人，与卡拉拉大理石的质地和混凝土的纹理相得益彰。

不言而喻，建筑师鲍赞巴克和克里斯蒂安·波尔坦斯基没能达到与毕赫葛里欧、马丁内斯或诺尔那样的默契程度。但前者对音乐学校的贡献令人景仰，颇有大家风范，尽管作品的场地位于保密区域，无法接近。他的设计与建筑师的意图不谋而合：创造强势且涵义深远的视觉艺术品，而非标准化的平庸之作。

雅克·马丁内斯为格拉斯法院中厅设计的60余米长的檐壁。
Jacques Martinez' 60-meter frieze for the law courts lobby in Grasse.

their meeting at art school and their shared commitments in 1968. But beyond this mutual allegiance — and perhaps preceding it — was a community of style that shines forth unmistakably when we compare certain of their works. The registers might be very different, but this is particularly true of what Buraglio did at Sainte Claire and Portzamparc's work for Strada Novissima.

Another accomplice was Jacques Martinez, who created the 60-meter frieze in the lobby of the Lawcourts in Grasse: another astute intuition, since the frieze, set up from floor level, encourages the visitor to "carry his head high". Paraphrasing the American architect Robert Venturi, Martinez has written that "Adding something in doesn't diminish the architecture." In 1990, eight years before Grasse, Xavier Girard wrote of the Martinez exhibition at the Château de Jau, "A lot of these canvases make me think of Christian de Portzamparc's Hautes Formes in Paris, with their paradoxical breaks set around a tall central axis...Martinez repeatedly uses what architects call overadjacency." The outcome, at Grasse, is a kind of nonchalance, sometimes almost gauche, but very tender and moving and in perfect harmony with the veining of the Carrara marble and the way the concrete has been textured.

Obviously we do not feel the complicity that exists with Buraglio, Martinez and Noël between the architect and Christian Boltanski. And yet the latter's contribution to the Conservatory is masterly and unforgettable, even if its setting is secret and inaccessible. In addition, it fits exactly with what the architect wanted: not standard visual art, yet something with power and meaning.

The permanent installation The Reserve of the Conservatory of Music (1991) is an integral part of the suites — Archives, Reserves and Dead Swiss — Boltanski began in the late 1980s which, like his Altars and Monuments are Lessons in Darkness.

In this work memory, oblivion and death repeatedly raise the question of identity, destiny and the razor's edge between truth and illusion. Ironic, grave and defying categorization, this is a work in which emotion cannot be separated from formal concerns. A work which, unquestionably, underpins the Conservatory with even greater intensity than all those tons of concrete. In turn issuer and executor of commissions, Portzamparc wove delicate webs of cooperation with his collaborators in an ongoing sequence of encounters, situations and affinities.

As an executor, he had to learn to deal with two sets of people: contracting authorities and clients. In the first case the learning process was long and arduous, in the second things went faster, more smoothly and more clearly. In both, however, it was up to the architect to grasp and

强健、引人入胜、但秘密且不可接近：克里斯蒂安·波尔坦斯基的《音乐学校收藏品》装置。

Masterly, striking, secret and inaccessible: Christian Boltanski in the Music Conservatory Reserves

《音乐学校收藏品》(1991)是永久性装置作品，成为一组艺术品的有机组成部分——《档案》,《收藏》,《死去的瑞士人》——波尔坦斯基在80年代后期着手这些创作，像《圣坛》和《墓碑》一样，它们都是"暗夜的训示"。

回首这些作品，泯灭与死亡的主题反复涉及身份、命运、以及真理与幻觉的鲜明分界等问题。在作品中，情感与对形式的关注合而为一，流露出讽刺韵味和庄重黯淡的气息，无法简单地归类。毋庸置疑，音乐城中的这项作品使建筑变得比那些厚重结实的混凝土更为密集浓烈、热情洋溢。身为任务的发布者和执行人，鲍赞巴克在众多参与者之间编织起密不透风的协作网络，处理此起彼伏的突发事件、复杂情况与种种关系。

作为执行者，他必须学会应付两类人：缔约决策者与业主。学会应对前一类人的过程漫长且艰辛，而后一种情况则显得较为快捷、顺利、清晰。但在两种情形下，建筑师都须担当起把握、解读服务对象的目标和设想，甚至超越他们的需要和预期的任务。

当有人质疑温斯顿·丘吉尔武断专横的决策时，他总是习惯于反驳，"你知道骆驼为何物么？它就是委员会设计出的兔子而已！"这比喻或许过于极端，但不可否认缔约决策者们总是义无反顾地将建筑师看作场地管理人：因此建造高于风格，最终产品重于辩证过程，标准先于欲求，规则范式优于随机应变。

年轻的建筑师也许会赢得音乐城竞赛，但他的设计团队仅有7人，这引起了拉维莱特委员会成员——由保罗·德鲁夫耶和塞吉·哥德堡领导——的猜疑与不信任。接二连三的冲突此起彼伏，持续数载，弗朗索瓦·密特朗总统和多位文化部长都涉足其中——杰克·朗，弗朗索瓦·莱奥塔尔，雅克·特布——工作时进时退，优

柔寡断，不止一次地停滞不前。但很多局外人却成为鲍赞巴克的拥趸——亨利·杜蒂耶，米歇尔·盖伊，奥利弗·梅西安，菲利浦·索莱尔——而在内部，鲍氏也赢得了项目领导者马克·布鲁兹和布吉特·迈格的支持，他们分别是音乐学校和音乐城两个子项的负责人。

一些人拥护"简单明了"的操作模式，另一些则希图雄心勃勃的计划能够引发出共鸣；而情况因鲍赞巴克对"出乎预料"、"奇异怪诞"和"不合逻辑"的偏爱变得愈加艰深晦涩。在鲍赞巴克的同道中人里，皮埃尔·布列兹是最可珍贵的支持者与知己："为了音乐城的椭圆形大厅设计，我和皮埃

音乐城模型展示。从左到右：克里斯蒂安·德·鲍赞巴克，弗朗索瓦·莱奥塔尔，塞吉·哥德堡和保罗·德鲁夫耶。
Presentation of the Cité de la Musique model. Left to right: Christian de Portzamparc, François Léotard, Serge Goldberg and Paul Delouvrier.

国立音乐舞蹈学校：彩色蜡笔画。
The National Music and Dance Conservatory: pastel on paper.

interpret what each was after, anticipating and even outstripping their needs and wishes.

Winston Churchill, when challenged over his rapid, peremptory decisions, had the habit of retorting, "Do you know what a camel is? It's a hare designed by a committee!" The comparison is perhaps a little extreme, but there can be no avoiding the fact that a contracting authority is, inevitably, a committee that sees the architect as a site manager: so the building takes priority over the style, the end-product over the dialectic, the specifications over the changes to be wrought, the standard over the desirable.

He might have taken out the Cité de la Musique competition, but this young architect with a team of only seven behind him generated mistrust and misgiving among the La Villette committee members, led by Paul Delouvrier and Serge Goldberg. The ensuing battle was to last for years and would involve, crucially, President François Mitterrand and a succession of Ministers of Culture — Jack Lang, François Léotard and Jacques Toubon — in a climate of to-ing and fro-ing, dithering and work more than once at a standstill. Outsiders pitched in on Portzamparc's behalf — Henri Dutilleux, Michel Guy, Olivier Messiaen, Philippe Sollers — while on the inside he gained the backing of Marc Bleuze and Brigitte Marger in charge, respectively of the project's two elements, the Conservatory and the Cité.

On one side then, the advocates of the "open and shut" approach, on the other those interested in the resonance of an ambitious proposal; a situation made all the more difficult by Portzamparc's recourse to the unexpected, the odd, the strange, the "illogical". Among the insiders, his most precious supporter was none other than Pierre Boulez: "For the oval room at the Cité de la Musique I put in a lot of work with Pierre Boulez. You don't want it rectangular? he asked me more than once. No, I said, oval. The thing is that acoustically speaking, ovals are terrible; there's always the danger of creating points of concentration. Sound concentrations are absolutely the worst, which is why there are no oval concert halls. To make matters worse, I wanted a high ceiling — fifteen meters —

尔·布列兹一道投入了大量精力，你不希望它是矩形么？皮埃尔不止一次地问过我。不，我说，椭圆形。关键在于，从声学角度讲，椭圆形态一无是处；它不可避免地具有出现声音集聚焦点的危险性。声音聚焦是大不韪，这也是为什么没有椭圆形音乐厅的缘由。更为糟糕的是，我还希望建造高大的顶棚——达 15 米——它将对早期反射声甚为不利。我们召开例会，布列兹最后一次建议我回归矩形，我坚持道：椭圆。他的回答是，我不理解这种形式，但我明白创造形式的欲望：这与音乐中的情形十分相仿，如果没有创造形式的愿望，你就将成为无头苍蝇。那么好，让我们动手吧，让我们着手设计椭圆音乐厅吧。”

布列兹对鲍赞巴克的礼遇令他回忆起对韦伯恩的印象：“多么倔强和固执，多么善于自我否定啊！一切均指向同一目标：找寻到一种语言，发掘出一种形式。”有趣的是，米歇尔·福柯对布列兹的评价如出一辙：“我们固执地认为，一种文化总会与价值观唇齿相依，而形式则位居次席；后者（指形式——译者注）变幻莫测，有时被摒弃一边，有时则被重新采纳；只有意义根深蒂固。然而，这是对随着形式的显隐出没而产生的惊世骇俗的震撼效果或憎恨厌恶之情的误解。它同样驱使我们误解如下观点——我们更加注意观看、表达、行为、思考的方式，而不是观看、表达、行为、思考本身。在西方，对形式的争斗较之对思想与价值的争论有过之而无不及。但在 20 世纪，情况变得异样起来：‘形式问题’——例如对形式系统的潜心研究——成为各式各样的道德对抗、美学争论与政治冲突的核心与关键点。”

我们应该为“委员会”作一些辩护，因为如果将创造当作游戏，那么管理就是追求更高层次利益的举动；如果说创造是信念与认同的问题，那么管理则代表现象的本质。

克里斯蒂安·德·鲍赞巴克，伯纳德·阿诺特和纽约路易威登大厦模型。
Christian de Portzamparc and Bernard Arnault with the modol of the LVMH tower in New York.

纽约路易威登大厦初步构思草图。
Preliminary drawing for the LVMH tower in New York.

which would be no use for the initial sound reflections. We had this big meeting, and for the last time Boulez asked me to come back to a rectangle, and I said: oval. His reply was, I don't understand the form, but I understand the will to form: it's the same in music, if there's no will to form you're headed nowhere. So okay, let's go with it, let's go with the oval."(*Voir-Écrire*, Paris, Calmann-Lévy)

Boulez's attitude to Portzamparc is reminiscent of his feelings about Webern: "What obstinacy, what self-denial! Everything focused on a single goal: finding a language, finding a form." Interestingly, Michel Foucault has the same thing to say about Boulez himself: "We are always ready to believe that a culture is more closely tied to its values than in its forms; that the latter can easily be modified, abandoned, taken up again; that only meaning is deep-rooted. But this is to misunderstand the sheer shock, or the hatred, forms can generate when they break down or make their appearance. This is to misunderstand that we are more concerned with ways of seeing, saying, doing and thinking than with what we see, think, say or do. The battle over form in the West has been as fierce, if not fiercer, that that over ideas and values. But things have gotten strange in the 20th century: the 'formal' as such, the business of working carefully on the system of forms, has become an issue — and the focus for all sorts of moral hostility, esthetic debate and political confrontations."

It has to be said in defense of "committees" that while creativity is play, administration is playing for high stakes; while creativity is a matter of convictions, administration is the nitty-gritty.

Sometimes, though, an exception pops up and a public-sector client turns out to have real commitment in both directions. One example was Michel Lombardini, from the City of Paris Real Estate Office, who had developed the habit of spotting youthful potential in the New Architecture Program competitions. He met Christian de Portzamparc, convinced him that he should enter competitions, observed him in action, talked with him and listened to him. Among the results were the Hautes Formes, the apartments on the Rue du Château des Rentiers, the Erik Satie Conservatory and Village 13.

Another vital acquaintance for Portzamparc was François Scheer, French ambassador to Germany at the time of the competition. Scheer had the gift of being able to communicate, in succinct,

事情也偶有峰回路转的时候，某些公共部门的业主间或在两个方向上都颇有作为。巴黎房地产委员会的米歇尔·隆巴蒂尼便堪称楷模，他在新建筑项目竞赛中注重提拔有为的年轻人。他会晤鲍赞巴克，说服他参与竞赛，积极关注事态发展，与他促膝攀谈，倾听他的心声。他们合作的成果包括欧风路住宅、位于霍尔城堡街的公寓，埃里克·萨蒂音乐学校，和 13 村项目等。

鲍赞巴克的另一位密友是弗朗索瓦·希尔，大使馆竞赛时，他出任法国驻德大使。希尔擅长交往，能够游刃有余地运用简明扼要、具体清晰的术语，即便在处理国家利益、外交策略、草案条约这类最为抽象的事务时，也显得行家里手。鲍赞巴克永不停歇地发掘这类目光深邃、忠诚不二的对话者身上的"附加价值"：脱身竞赛之外，情况就变得柳暗花明起来，业主主动选择了你，因此"赏心悦目"成为关键。比如伯纳德·阿诺特和吉尔伯特·柯斯蒂斯就属于这类情形，鲍赞巴克在一次宴请克林顿总统的总理级宴会中，和一次私人聚餐上分别结识了他们：桌旁的闲聊最终促成了纽约路易威登大厦和巴黎的波布尔咖啡馆两个项目。

诚然，鲍赞巴克此前已经接触过路易威登公司的高层，但没有任何决定板上钉钉。总理宴会成为千载难逢的机遇，进而促成鲍氏的纽约之行，他踏勘现场，为公司展示初步构思草图，制作模型，伯纳德·阿诺特还拜访了他的工作室——自此以后，他一直毫不懈怠地关注这个项目。阿诺特的团队向纽约展示的与其说是力量，倒不如说是创造力；而鲍赞巴克正是及时雨般的福将。阿诺特自始至终对他偏爱有佳，从基础铺设到上层建筑，事无巨细，都表现出经久不衰的睿智与积极态度，鲍赞巴克对此喜出望外。与吉尔伯特·柯斯蒂斯的奇遇则显得既相似却又有所不同。餐后，两人促膝探讨方案。吉尔伯特的兄弟让·路易斯已经邀请菲利普·斯塔克设计柯斯蒂斯咖啡厅，这次轮到吉尔伯特出山了。他在蓬皮杜中心旁边获得了四座店铺，包括一座被公寓楼入口一分为二的咖啡厅：他对咖啡厅有自己的心得——尤其是这座与巨大的建筑物相邻接、被视为"附庸"的咖啡厅。鲍赞巴克面露难色，他刚刚在音乐城竞赛中金榜题名，将自己的团队扩充至 25 人，并搬迁到欧德路。伊丽莎白鼓励他，力劝他回心转意。鲍氏很会揣摩这位潜在业主的种种需要，朝夕之间就规划出一座咖啡厅，一座所有人都能各得其所的咖啡厅：观看与被观看，独处，私人会晤、朋友聚会或集体活动，样样不缺。他创造出移步换景的空间、视角、走廊，并在某个星期五将方案提交给吉尔伯特·柯斯蒂斯。第二周的星期一，潜在的业主就升级为名副其实的业主了。

对话、讨论、交换、偶遇：这些要素成为建筑的基本组成部分，尤其是在沟通与交谈被信息通讯所取代的危机重重的时代里。

整整 2500 年前，埃斯库罗斯在《阿伽门农》中坦言，"听着。我知道这件事冗长而艰难，但是，恳请您凝神倾听，否则我只有随风仙逝。"

克里斯蒂安·德·鲍赞巴克和吉尔伯特·柯斯蒂斯。
Christian de Portzamparc and Gilbert Costes.

波布尔咖啡馆初步构思草图。
Preliminary drawing for the Café Beaubourg.

by LVMH senior management, but nothing had been decided. The Prime Ministerial dinner was a significant encounter, leading to a Concorde flight to New York to check out the site, presentation of initial sketches to the company, preparation of models and a visit to the Workshop by Bernard Arnault, whose involvement was unflagging from there on in. The challenge for Arnault was to show New York not so much the power as the creativity of his group; and for him Portzamparc was the man of the hour. Arnault stuck to him throughout, from groundwork to building, displaying an unfailing acuity and reactivity that Portzamparc found nothing less than delightful. The venture with Gilbert Costes was at once different and similar. After dinner the two discussed a project together. Gilbert's brother Jean-Louis had already had his Café Costes designed by Philippe Starck and now it was Gilbert's turn. Next to the Centre Pompidou he had acquired four shops, including a café, that were split in two by the entrance to the apartment block: and he had his own ideas of what a cafe ought to be — especially this one, envisaged as an "annex" to its enormous neighbor. Portzamparc, who had just won the Cité de la Musique competition and was moving into the Rue de l'Aude with an expanded team of 25, hesitated. Elizabeth encouraged him, urged him on. He had a perfect grasp of what the potential client wanted and very quickly came up with a project covering everything people could want from a cafe: to see and be seen, to be alone, to chat privately or with friends, to arrive as a group, and so on. He created all sorts of spaces, vistas, passageways, and submitted the project to Gilbert Costes one Friday. By the following Monday the potential client was — a client.

concrete terms, even the most abstract notions to do with the State, diplomacy, protocol and so on. Portzamparc insists endlessly on the "added value" aspect of this kind of clear-sighted, committed interlocutor: the kind of situation you find more easily, of course, outside the competition situation, when the client has chosen you and the "desirable" has priority. Such was the case with Bernard Arnault and Gilbert Costes, whom he met, respectively, at a Prime Ministerial dinner in honor of President Clinton and at a private dinner party: tabletalk eventually led to the LVMH tower in New York and the Café Beaubourg in Paris.

True, Portzamparc had already been approached

Dialogue, discussion, exchange, encounter: these make up a fundamental component of architecture, especially in an age in which communication has dangerously supplanted conversation.

All of 2500 years ago Aeschylus, in Agamemnon, said, "Listen. I know it's a long and difficult business, but listen to me or I'll die."

Exposition Scènes d'Atelier au centre Georges Pompidou, en 1996
Scènes d'Atelier Centre Georges Pompidou exhibition, 1996

DIRECTEURS DE STUDIO
STUDIO DIRECTORS
Bertrand Beau
Bruno Durbecq
Benoit Juret
Marie-Elisabeth Nicoleau
Etienne Pierres

CHEFS DE PROJETS
PROJECT MANAGERS
Céline Barda
Frédéric Binet
Barbara Bottet
Jean Daniel Boyé
Eric Brunel
Karol Claverie
Christophe Eschapasse
Léa Xu

COLLABORATEURS
ASSOCIATES
Pascal Boutet
Duccio Cardelli
Florence Clausel
Chantal Crouan
Anne-Catherine Detry
Constantin Döhler
Nanda Eskes
Matthieu Falliu
Nicolas Gicquel
Frédéric Lagardère
Cécile Loe-Mie
Renaud Magnaval
Gaëlle Maidon
Ana Paula Pontes
Odile Pornin
Isabelle Ragot
Jeanne Ravet
Daniel Roméo
Arianne Rupp
Jean-Philippe Salles
Burkhardt Schiller
Amalia Sevilla
Urban Stirnberg

ADMINISTRATION
ADMINISTRATION
Françoise Casanova
Marie-France Boucher
Laetitia Viciot

GESTION
MANAGEMENT
Marie-Thérèse Glénisson
Mohamed Arris
Robert Lamy

COMMUNICATION
COMMUNICATION
Gaëlle Martin
Gabriella Wilson Sadler

IMAGE
VISUALS
Jean-Charles Chaulet
Thierry Damez-Fontaine
Michael Kaplan
Abderhaim Nouaiti

ONT TRAVAILLÉ 1980's
WORKED IN THE 1980's
Olivier Blaise
Stéphane Boesse
Frédéric Borel
François Chochon
Nadine Clément
John Coyle
John Curran
Marianne Daviaud
Corinne do Nascimento
Saskia Fokkema
Maryvette Georges
Eamon Gogarty
Alexia Guerin
Paul Guilleminot
Catherine Hervé
Bruno Huber
Pascal Joncour
Florent Léonhardt
Jean-Sébastien Malliard
Sam Mays
David Mc Nulty
Petr Opélik
Ségolène de Roquemaurel
Muriel de Rouyn
Richard Scoffier
Odile Van der Brock

ONT TRAVAILLÉ 1990's
WORKED IN THE 1990's
Aro Andonissamy
Christine Audouin Hein
Pierre Auffan
François Barberot
Matthieu Belin
Wilfrid Bellecour
Valérie Berger
Michel Bogar
Jean-Pierre Buisson
Francisco Carriba Anta
Ellen Cassily
Bob Clément
Kaan Coskun
Christophe Evenard
Maria-Teresa Fernando
Denise François
Gilles Greffier
Régis Gullon
Nicolas Jancovic
Jacqueline Jourdain
Min-A Jung
Julie Howard
Nicolas Karmochkine
Juchini Kato
Dominique Lim
Annick Liot
Alexi Lorch
Nicola Marchi
Philippe Michel
Lene Mizrahi Mirdal
Vania Nalin Uehara
Isabelle N'Guyen
Rémy Oliel
Laurent Pierre
Cristina Pilara
Cyrille Pinelli
Michel Plantain
Otavio Ribeiro
Sandra Roten
Anne-Sophie Rouaffaert
Pascal Sellam
Hyun Jung Song
Olivier Souquet
Marianne Swierczewski
André Terzibachian
Anne Traband
Pierre Van den Berg
Jean-Marc Venne
Thanh Vuong

COMPLICES
CONTRIBUTORS
Partenaires/Partners
Gary Handel
Antoine Felix-Faure
Philippe Macary
L.A. Rangel
Acousticiens/Acousticians
Jean Paul Lamoureux,
José Augusto Nepomulceno,
Xu Ya Ying
Économie/Economists
Pascal Ambiaud (dal),
Henrique de Arago
(engineering),
David Chalverat (gvi),
Francis Christol (atec)
Jean Luc Dubuisson (atec),
Jea Louis Laussine (atec)
Maquettistes/Model makers
Catherine Claden,
Alain Hugon
Photographes/Photographers
Nicolas Borel,
Gitty Darugar,
Deidi van Schaewen
Scénographes/Scenography
Jacques Dubreuil,
Michel Fayet
Ismaël Solé
Sculpteur/Sculptor
Martin Wallace
Structure/Structure
Joseph Attias,-
Jean-Bernard Datry (setec),
Bruno Cantarini,
Carlos Fragelli
BET Façade/BET Frontage
Michel Mourot,
Castaing (thales),
Bob R. Heintges,
Bertrand Toussaint (rfr)

Crédits photographiques
Photographic credits

Toutes photographies, Nicolas Borel, sauf-:
All photographs, Nicolas Borel, except:

- François-Xavier Bouchard: p. 296
- Alain Dagbert: p.28
- Gitty Darugar: p. 29 bas, p.225 bas dt, p.226
- D.R.: p.16, 17, 32, 33, 34, 36, 37, 41, 117 haut dt, 131 haut dt, 291 centre et bas, 294, 295, 296, 298, 301
- Magnum/photos: p. 266
- Steve Murez: p. 34 bas, p.236 bas centre
- Parc de la Villette. Fonds S.E.M.U.I.: p.48
- Luc Perenom: p.11
- Étienne Pierres: p. 158
- Deidi von Schaewen: p. 86, 92 bas, 290

Imprimé par Graphiche Zanini
Bologne - Italie
Septembre 2003